2nd International Conference on Smart Materials Applications (ICSMA 2019)

AA006636

IOP Conference Series: Materials Science and Engineering
Volume 559

Tokyo, Japan
19 – 22 January 2019

ISBN: 978-1-5108-9093-0
ISSN: 1757-8981

Printed from e-media with permission by:

Curran Associates, Inc.
57 Morehouse Lane
Red Hook, NY 12571

Some format issues inherent in the e-media version may also appear in this print version.

This work is licensed under a Creative Commons Attribution 3.0 International Licence. Licence details: http://creativecommons.org/licenses/by/3.0/.

No changes have been made to the content of these proceedings. There may be changes to pagination and minor adjustments for aesthetics.

Printed by Curran Associates, Inc. (2019)

For permission requests, please contact the Institute of Physics at the address below.

Institute of Physics
Dirac House, Temple Back
Bristol BS1 6BE UK

Phone: 44 1 17 929 7481
Fax: 44 1 17 920 0979

techtracking@iop.org

Additional copies of this publication are available from:

Curran Associates, Inc.
57 Morehouse Lane
Red Hook, NY 12571 USA
Phone: 845-758-0400
Fax: 845-758-2633
Email: curran@proceedings.com
Web: www.proceedings.com

Table of contents

Volume 559

2019 International Conference on Smart Materials Applications

19–22 January 2019, Tokyo, Japan

Accepted papers received: 21 May 2019

Preface

Papers

Material Science and Engineering

Industry and Chemical Technology

IOP Conf. Series: Materials Science and Engineering **559** (2019) 011001 doi:10.1088/1757-899X/559/1/011001

Preface

ICSMA 2019 aims at creating a platform for discussing the research issues, opportunities and challenges of Smart Materials Applications. The ever-changing scope of Smart Materials Applications brings along new research questions resulting in a need to stimulate better awareness of this very important research field, especially among young researchers. For us, ICSMA2019 promises a brighter picture for Smart Materials Applications while the support received and the enthusiasm witnessed has truly exceeded our expectations.

Through our efforts, we tried to spread the news about ICSMA 2019 to as many relevant researchers of the world as possible. Towards that we were blessed with an energetic team, the members of which worked side-by-side which resulted in optimization of coordination among all.

The present proceedings includes the selected papers of the 2nd International Conference on Smart Materials Applications (ICSMA 2019), which was Tokyo, Japan during January 19-22, 2019. The ICSMA 2019 as a global academic event focused on smart materials applications. The conference brought together the latest discoveries of researchers with the experiences of leading industry experts over the world. The ICSMA 2019 presented the innovations and new technologies that will charter the path to the smart materials applications of the future. Some main key topics included in the proceedings are as following:

- Materials Science and Engineering
- Materials Properties, Measuring Methods and Applications
- Methodology of Research and Analysis and Modeling
- Materials Manufacturing and Processing

Much of the credit of the success of the conference is due to topic coordinators who have devoted their expertise and experience in promoting and in general co-ordination of the activities for the organization and operation of the conference. The coordinators of various session topics have devoted a considerable time and energy in soliciting papers from relevant researchers for presentation at the conference.

We would like to thank everyone who was part of this conference and would like to see you again at ICSMA 2020.

Conference Chairs

On behalf of the ICSMA Conference Committee

Prof. Xiaohong Zhu, Sichuan University, China

January 19-22, 2019

Content from this work may be used under the terms of the Creative Commons Attribution 3.0 licence. Any further distribution of this work must maintain attribution to the author(s) and the title of the work, journal citation and DOI.
Published under licence by IOP Publishing Ltd

IOP Conf. Series: Materials Science and Engineering **559** (2019) 011002 doi:10.1088/1757-899X/559/1/011002

CONFERENCE COMMITTEES

Conference Chairs

Prof. Kazuo Umemura, Tokyo University of Science, Japan
Prof. Xiaohong Zhu, Sichuan University, China

Program Committee Chairs

Prof. Shu YIN, Tohoku University, Japan
Prof. Yu-Liang Chen, National Defense University, Taiwan
Assoc. Prof. Teik-Cheng Lim, Singapore University of Social Sciences, Singapore

Technical Committee Chairs

Prof. Stoyko Fakirov, University of Auckland, New Zealand
Prof. Hyoung-Joon Jin, Inha University, South Korea
Prof. Hiroshi Uechi, Osaka Gakuin University, Japan
Prof. Vikas Rastogi, Delhi Technological University, Indian
Prof. Pramuan Tangboriboonrat, Mahidol University, Thailand
Assoc. Prof. Jose Isagani B. Janairo, De La Salle University, Philippines
Assoc. Prof. Abdul Hai Alami, University of Sharjah, United Arab Emirates
Asst. Prof. Dr. Francesca Scalisi, University of Palermo, Italy
Asst. Prof. Shadi Yasin, American University of the Middle East, Kuwait
Dr. Alicja Klimkowicz, Shibaura Institute of Technology, Japan
Dr.Tomáš Homola, Masaryk University, Czech Republic
Dr. HUYNH, TRONG-PHUOC, Cantho University (CTU), Vietnam
Dr. Sudip Kumar Sinha, National Institute of Technology Raipur, India
Dr.Risti Suryantari, Parahyangan Catholic University, Indonesia
Dr. Koorosh Gharehbaghi, RMIT University, Australia
Karel Raz, University of West Bohemia, Czech Republic
Dr.Kittipoom Rodsin, King Mongkut's University of Technology North Bangkok,Thailand
Associate Professor Hosein Naderpour, Semnan University,Iran

Content from this work may be used under the terms of the Creative Commons Attribution 3.0 licence. Any further distribution of this work must maintain attribution to the author(s) and the title of the work, journal citation and DOI.
Published under licence by IOP Publishing Ltd

Peer review statement

All papers published in this volume of *IOP Conference Series: Materials Science and Engineering* have been peer reviewed through processes administered by the proceedings Editors. Reviews were conducted by expert referees to the professional and scientific standards expected of a proceedings journal published by IOP Publishing.

Content from this work may be used under the terms of the Creative Commons Attribution 3.0 licence. Any further distribution of this work must maintain attribution to the author(s) and the title of the work, journal citation and DOI.
Published under licence by IOP Publishing Ltd

IOP Conf. Series: Materials Science and Engineering **559** (2019) 012001 doi:10.1088/1757-899X/559/1/012001

Dynamic Mechanical Properties of Magnetorheological Elastomers: Experiment and Modelling

G H Zhu[1], Z G Li[1], Y P Xiong[2], L Xiao[1] and M Li[1]

[1] Department of Mechanics, Xi'an University of Science and Technology, Xi'an, China

[2] Faculty of Engineering and the Environment, University of Southampton, Southampton, UK

E-mail: zhuguanghong@xust.edu.cn

Abstract. As a kind of smart material, magnetorheological elastomer (MRE) is composed of magnetizable particles dispersed in an elastomer matrix. Because adjusting an external magnetic field can continuously, rapidly and reversibly control the dynamic properties, there has been increasing research on MRE for mitigation of unwanted vibrations. In this paper, the dynamic mechanical analysis tests were performed to investigate the influence of strain, frequency and magnetic field on the controllable dynamic properties. The storage modulus and loss modulus of MRE were analyzed with frequencies from 1 to 50 Hz, strain amplitudes from 1 to 6% and magnetic field intensities from 0 to 500 mT. The results show that the storage modulus decreases with increasing strain amplitude, increases with frequency and magnetic strength, and remains constant when the magnetic saturation occurs. Furthermore, the dependence of loss modulus on frequency lies on the matrix material of MRE. Based on the experimental results, a fractional derivative model was developed to describe the viscoelastic properties and the controllable dynamic properties. The comprehensive study of mechanical property characterisation is a guarantee for constitutive models to accurately describe the dynamic behaviour of MRE, which is an essential step towards the application for vibration control.

1. Introduction

As a composite material, MRE consists of magnetisable particles and non-magnetic matrix. MRE can be classified into two types [1]: anisotropic MRE which have a directed particle orientation attributed to the application of magnetic field during the solidification, and isotropic MRE which can be characterized by a random distribution of magnetic particles as the curing progress is without magnetic field. When a magnetic field is applied, the magnetic particles tend to align along the magnetic field direction, as a result, the composite exhibits field dependent properties within a pre-yield regime [2]. Because the dynamic stiffness can be selectively tuned by controlling magnetic fields [3] and finely tailored by adjusting the isotropic/anisotropic ratio [4], MRE is highly suitable for semi-active vibration control systems. Due to the combination of the versatility of active control and the reliability of passive control, semi-active (adaptive-passive) vibration control devises have attracted considerable intention over the past decades, and increasing effort has been devoted to the possibility of incorporating smart materials into semi-active vibration control devices [5]. Among these suited smart materials, MRE is distinctive due to thc controllable rheological property by adjusting an external

Content from this work may be used under the terms of the Creative Commons Attribution 3.0 licence. Any further distribution of this work must maintain attribution to the author(s) and the title of the work, journal citation and DOI.
Published under licence by IOP Publishing Ltd

magnetic field. There has been increasing research on MRE based devices to commercialize and industrialize its application in varies fields, such as thc automotive industry and civil engineering [6,7]. Among all the influence factors on the dynamic properties of MRE, temperature can lead to transition behaviour at about 50°C. The storage modulus exhibits two different trends with the temperature variation: firstly decreases rapidly and then increases slightly with increasing temperature [8]. In this paper, an experimental research on mechanical property characterisation of MRE was carried out at a stable temperature to investigate the influence of strain, frequency and magnetic field on the controllable dynamic properties. This experimental study will be an important basis for the establishment of a fractional derivative model, which is capable to describe the viscoelastic properties and the controllable dynamic properties with strain, frequency and magnetic field strength as independently continuous variables. Accurate models are indispensable for effective dynamical analysis to realise the application of MRE, therefore, dynamical modelling is important as the basis for the progress of accomplishing the application of MRE.

2. Background theory

When sinusoidal loads are applied to MRE materials, because of viscoelasticity the response will not be instantaneous. The resulting stress will lag behind the input load by an angle φ called loss angle whose range is $0<\varphi<90°$. In viscoelastic materials, some of the deformation energy can be stored and recovered, whilst the remainder is dissipated as heat during each cycle. The storage modulus M' represents the ability of viscoelastic material to store the energy due to deformation, which contributes to the material stiffness. The loss modulus M'' indicates the ability of viscoelastic material to dissipate the energy of deformation. They can be defined as:

$$\sigma = \varepsilon_0 (M' \sin\omega t + M'' \cos\omega t) \tag{1}$$

where M can be either the shear modulus G or the Young's modulus E. It is usually convenient to express the modulus as a complex quantity. Then the dependence of the in-phase and out-of-phase stress on the strain can be presented using the complex modulus M^*:

$$M^* = M' + iM'' . \tag{2}$$

The ratio between the loss and storage moduli is another widely used term for viscoelastic materials:

$$\tan\varphi = M''/M' \tag{3}$$

where $\tan\varphi$ is called the loss factor, whic h can be used for describing the efficiency of damping caused by the viscoelastic material.

According to the Dynamic Mechanical Analysis directions in this study, the complex modulus M^* can be obtained by calculating the ratio of the stress range to the strain range. The loss angle φ is relevant to the energy dissipation per volume within an oscillatory cycle, which is the area enclosed by the hysteresis loop and can be calculated. The storage modulus M', loss modulus M'' and loss angle φ can be defined as:

$$\begin{aligned} M' &= M^* \cos\varphi \\ M'' &= M^* \sin\varphi \end{aligned} \qquad \varphi = \arcsin\frac{E_{loop}}{\pi A_{strain} A_{stress}} \tag{4}$$

where A_{strain} stands for the strain amplitude, A_{stress} stands for the stress amplitude and E_{loop} is the energy enclosed by the hysteresis loop where the numerical integration can reduce the effects of measurement noise and take into account waveform distortion.

A fractional element is introduced to describe the characteristics of viscoelastic materials, which perform between viscous and elastic behavior [9]. The relationship between stress and strain can be expressed in the following form,

$$\tau_v = G_1 \tau_0^\alpha D^\alpha \left[\gamma_v(t)\right] \quad 0<\alpha<1 \tag{5}$$

where G_1, τ_v and γ_v are the shear modulus, the shear stress and the shear strain of fractional element, respectively, τ_0 denotes the relaxation time constant, D denotes the derivative operator, α is a fractional order. In the extreme cases of α=1 and α=0, the fractional element turns into a Newton dashpot and a Hook spring, respectively.

Based on the experimental results of dynamic mechanical analysis tests, a fractional derivative model is developed as shown in Figure 1, and the Fouriour transform is used to obtain the storage modulus G' and the loss modulus G''

$$G'(\omega)=G_m+\frac{G_0G_1^2\tau_0^{2\alpha}\omega^{2\alpha}+G_0^2G_1\tau_0^{\alpha}\omega^{\alpha}\cos\frac{\alpha\pi}{2}}{G_0^2+G_1^2\tau_0^{2\alpha}\omega^{2\alpha}}\quad G''(\omega)=\frac{G_0^2G_1\tau_0^{\alpha}\omega^{\alpha}\sin\frac{\alpha\pi}{2}}{G_0^2+G_1^2\tau_0^{2\alpha}\omega^{2\alpha}} \tag{6}$$

where G is the equivalent shear modulus, G_0 and G_m are the shear moduli of the Hook spring and the nonlinear spring, respectively, ω is the angular frequency of the sinusoidal load.

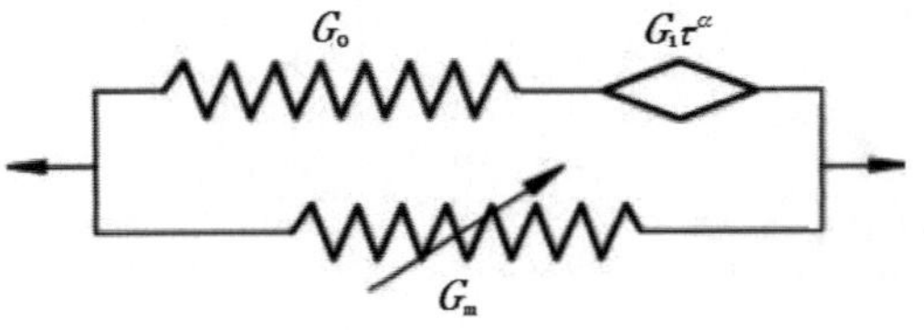

Figure 1. Fractional derivative model with a Hook spring in series with a fractional element, and then in parallel with a nonlinear spring.

3. Experimental setup

The anisotropic MRE samples were comprised of the silicone rubber (Wacker Chemie AG, Germany) and micron-sized iron particles (Sigma-Aldrich, US). The simplified manufacturing procedure can be illustrated by three steps: firstly Elastosil A and Elastosil B were mixed with a ratio of 10:1 by volume, and then the carbonyl iron powders sized up to 9 μm were added with a particle volume concentration of 30%, which is considered to be able to generate the maximum MR effect [10]; secondly, the mixture was blended thoroughly and placed in a vacuum chamber for 20 minutes to remove air bubbles trapped inside; finally the mixture was put into square aluminium moulds 21.8 × 21.8 × 6.5 mm³ and cured for 16 hours at room temperature to solidify with an external magnetic field of 290 mT produced by the cylindrical grade N42 neodymium permanent magnets (E-magnets, UK).

According to the BS ISO 4664-1:2011 for shear modes, the DMA tests of MRE samples were performed with Instron E1000 Electro plus, which is able to apply and control the harmonic shear strain, as shown in Figure 2. The influence of magnetic field on the dynamic properties was obtained by the use of cylindrical grade N42 neodymium permanent magnets (E-magnets, UK), meanwhile the force and displacement were tracked by Instron E1000 Electro plus during tests. The dynamic properties of MRE were measured with frequencies from 1 Hz to 50 Hz, strain amplitudes from 1% to 5%, and magnetic field intensities from 0 to 500 mT. And all the sets of tests were carried out at room temperature (about 25°C) with three pairs of MRE samples independently.

4. Mechanical properties of MRE

Based on the hysteresis loops of anisotropic MRE in Figure 3, it can be seen that the slope of hysteresis loop decreases with increasing strain amplitudes, indicating that the storage modulus decreases with the increasing strain amplitude of the shown applied harmonic load.

It can be seen from Figures 4 that the storage modulus and loss modulus increases with magnetic field strength at various excitation frequencies, and when magnetic saturation occurs these moduli will remain constant with the maximum magnetic force between magnetisable particles. Most experimental research on MRE shows that higher than 500 mT the increase of storage and loss moduli gradually slows down with magnetic field strength, and above about 800 mT magnetic saturation occurs in the composite there is no more increase of storage and loss moduli with further increasing magnetic field intensities [2]. It is also obvious that the storage modulus and loss modulus decrease with the strain amplitude, because when the strain amplitude is 1% the two moduli are higher than them when the strain amplitude is 5%.

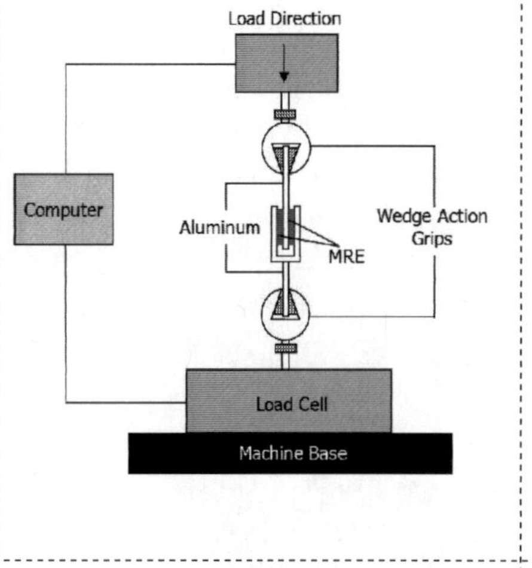

Figure 2. Experimental setup for MRE samples in shear mode.

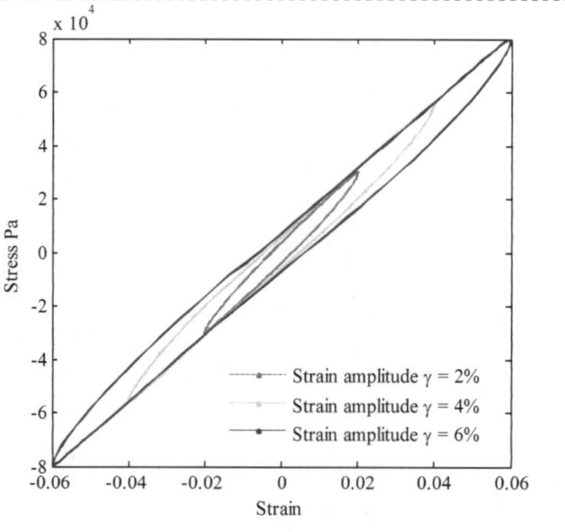

Figure 3. Dependence of hysteresis loops on strain amplitude.

Figure 5 indicates that the storage modulus increases with frequency, meanwhile the loss modulus initially increases with driving frequency, and then reaches a maximum around 10Hz and stops to decrease with further increasing excitation frequency. Because the loss modulus of silicone rubber shows the same change with frequencies, the matrix material determines the dependence of loss modulus on frequency. Furthermore, the goodness of fit proves the effectiveness of the fractional derivative model. With the use of genetic algorithm (GA) in Matlab optimization toolbox, the parameters of the fractional derivative model can be identified by minimizing the sum of square of error between the experimental data and predicted results of storage modulus and loss modulus.

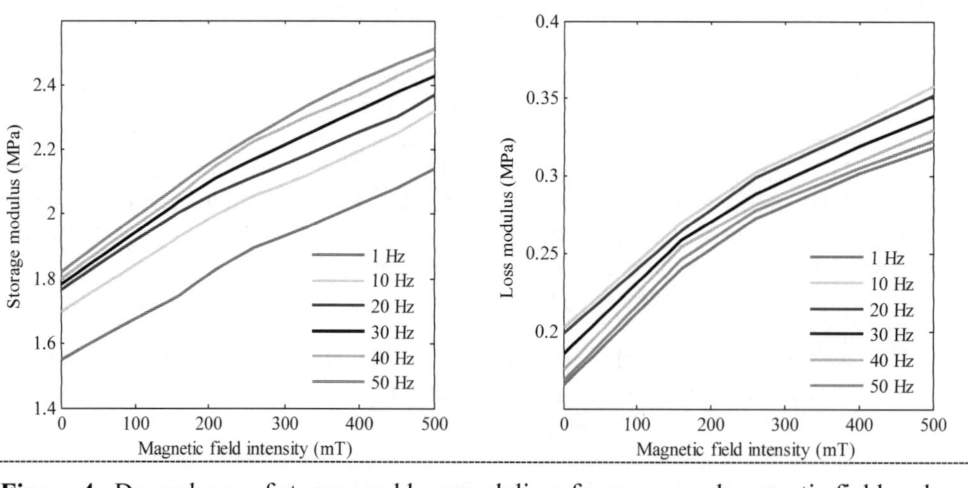

Figure 4. Dependence of storage and loss moduli on frequency and magnetic field under strain amplitude of 1%.

5. Conclusions

This study experimentally investigated the dependence of dynamic properties on frequency, strain and magnetic field for MRE in shear mode. The primary experimental results are as follows.

1. Both the storage modulus and loss modulus of MRE decrease with increasing strain amplitudes.

2. Both the storage modulus and loss modulus of MRE increase with magnetic flux density and maintain constant values when the magnetic saturation occurs.
3. The storage modulus of MRE increases as the excitation frequency increases, but the dependence of loss modulus on frequency is determined by the matrix material of MRE: in this study the loss modulus initially increases with increasing excitation frequency (<10Hz) up to a maximum value and then it decreases slightly with further increasing excitation frequency.

And finally, a fractional derivative model was developed to describe the dynamic properties of MRE, and the model prediction exhibited good coincidence with experimental results.

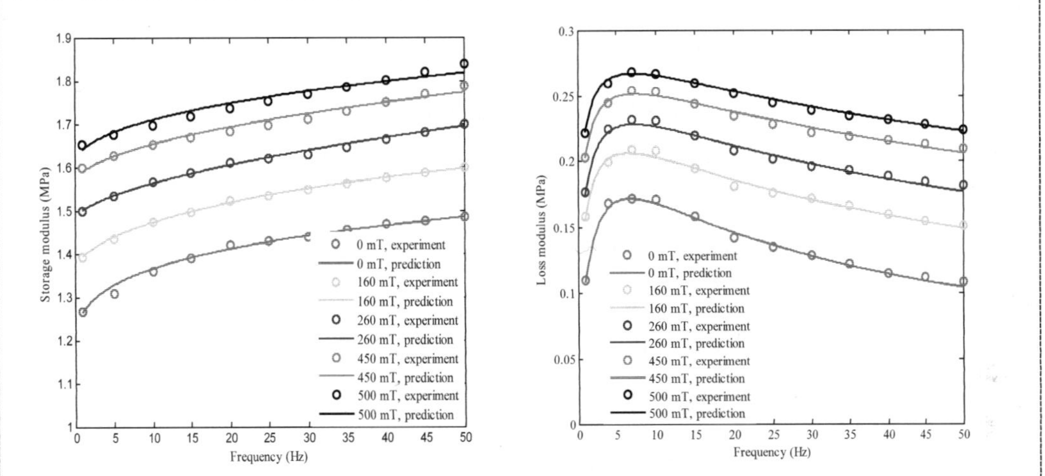

Figure 5. Comparison between measured data and predicted results of storage and loss moduli under strain amplitude of 5%.

6. References

[1] Kaleta J, Krolewicz M and Lewandowski D 2011 Magnetorheological properties of anisotropic and isotropic magnetorheological composites with thermoplastic elastomer matrices *Smart Materials and Structures* **20(8)** 085006

[2] Bose H and Roder R 2009 Magnetorheological Elastomers with High Variability of their Mechanical Properties *Journal of Physics: Conference Series* **149(1)** 012090

[3] Li Y, Li J, Li W and Du H 2014 A state-of-the-art review on magnetorheological elastomer devices *Smart Materials and Structures* **23(12)** 123001

[4] Sapouna K, Xiong Y P and Shenoi R A 2017 Dynamic mechanical properties of isotropic/anisotropic silicon magnetorheological elastomer composites *Smart Materials and Structures* **26(11)** 115010

[5] Ibrahim R A 2008 Resent advances in nonlinear passive vibration isolators *Journal of Sound and Vibration* **314(3)** pp 371-452

[6] Du H, Li W and Zhang N 2011 Semi-active variable stiffness vibration control of vehicle seat suspension using an MR elastomer isolator *Smart materials and structures* **20(10)** 105003

[7] Li Y, Li J, Li W and Samali B 2013 Development and characterization of a magnetorheological elastomer based adaptive seismic isolator *Smart Materials and Structures* **22(3)** 035005

[8] Wan Y, Xiong Y P and Zhang S 2018 Temperature dependent dynamic mechanical properties of Magnetorheological elastomers: experiment and modeling *Composite Structures* DOI: 10.1016/j.compstruct.2018.04.010

[9] Sasso M, Palmieri G and Amodio D 2011 Application of fractional derivative models in linear viscoelastic problems *Mechanics of Time-Dependent Materials* **15(4)** pp 367-387

[10] Davis L C 1999 Model of magnetorheological elastomers *Journal of Applied Physics* **85(6)** pp 3348-3351

Development of High Performance Grouting Method for PSC Strands

In-Seok Yoon[1], Joo-Hwan Kim[1], Min-Jin Jeon[1] and Seoung-Hwan Jang[1]

[1]Dept. of Const. info. Eng. Induk University, Seoul01878, Republic of Korea

Email: isyoon@induk.ac.kr

Abstract. Prestressed concrete is very vulnerable to corrosion of strand because it has already been subjected to high tensile stress. The stability and durability of PSC have a very important relationship. In this study, many grouting specimens was made and the grouting performance used in the post tensioning construction fields were evaluated by the electrochemical method and weight loss measurement. The corrosion test was carried out in accordance with modification based on PTI M55.1-12 specification. Corrosion was activated by two types: corrosion by chloride ion (Cl^-), and corrosion by chloride ion and dissolved oxygen (Cl^- + DO). The results showed that polymer is useful to maintain structural stability by reducing deterioration due to strands corrosion in PSC

1. Introduction

Corrosion-related failures of strand have been reported in the late 1990's, USA. Strand failure was also examined in concrete viaduct in Republic of Korea as well. The normal causes have been attributed to development of bleed water during grouting and subsequent re-absorption or evaporation from the grout. Poor grouting during construction were of concern[1-3]. Corrosion failure of post-tensioned strands has been also identified to be related to grout segregation in Europe. Common main causes of premature prestressing strand corrosion are known as chloride or sulfate ions, microcracks, and voids due to insufficient cement grout, oxygen, and humidity[4]. The cement grout filled in the duct of bonded tendon forms a highly alkaline environment and corrosion resistant oxide layer on the surface of the prestressing strand. When the cement grout is exposed to salt water or CO_2 gas, corrosion can be initiated by chloride ions penetration or carbonation, as shown in Figure 1. After corrosion initiation, the rate of corrosion is governed by various factors such as oxygen inflow, water content, electrical resistance, and pH of cement grout[5]. The tendon at ions, galvanic interaction with the metal ducts and anchorages and possible enhanced macrocell corrosion.

According to the presence of corrosion of the prestressing tendon, the life cycle stages of prestressed concrete members have three life cycle stages, (1) crack initiation stage, (2) slow crack propagation stage, and (3) fast crack propagation stage, which induces a rapid brittle failure of prestressing strand by stress corrosion cracking. This is due to the stress corrosion phenomena and hydrogen embrittlement, which deteriorate metal structures. The higher the strength of prestressing strand, the more vulnerable to the brittle failure induced by hydrogen embrittlement. Higher degree corrosion coupled with higher levels of prestress leads to higher losses of prestress and this was accounted for as an additional loss at the design stage[6].

Based on the technical background above, strand corrosion is very important for durable prestressed concrete system, nevertheless, it has been rare to deal with this issue. This study is, therefore, devoted to examine the grouting performance of corrosion protective systems widely used in the multi-strand post-tension construction

Content from this work may be used under the terms of the Creative Commons Attribution 3.0 licence. Any further distribution of this work must maintain attribution to the author(s) and the title of the work, journal citation and DOI.
Published under licence by IOP Publishing Ltd

fields. This work can provide us with how to make high performance grouting to make sure the best performance of corrosion protective systems.

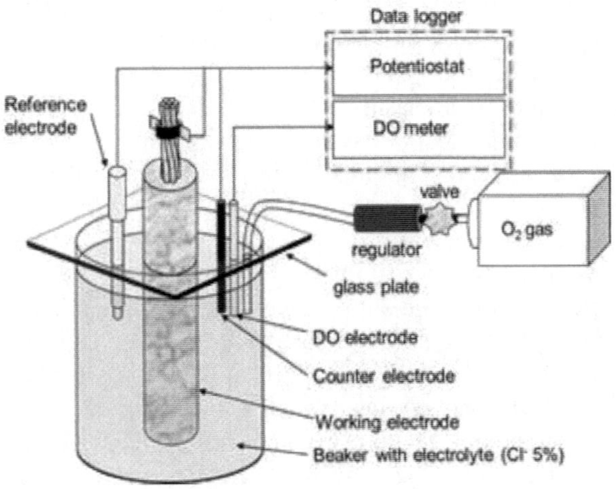

Figure 1. Corrosion cell

2. Experiment overview

2.1 The Creation of Test Piece

Testing samples were casted with a circular cylinder pipe, placed on strands in the center. Seven strands with a nominal diameter of 16 mm were used. Mixing table is as shown in Table 1. Experimental variation is an amount of mineral admixture, anti-corrosion and polymer. Grouting was cured for 28 days, and then corrosion testing was carried out. Samples were sealed with epoxy at both ends.

Table 1. Grouting mixing table.

Mixture symbol	Cement	*Water*	Blast furnace slag	Fly ash	Polymer	*Anti-corrosion*
P0A0	1	0.27	0.4	0.1	0	0
P7A0	1	0.27	0.4	0.1	0.07	0
P15A0	1	0.27	0.4	0.1	0.15	0
P0A5	1	0.27	0.4	0.1	0	0.05
P0A10	1	0.27	0.4	0.1	0	0.1

P[1] **0**[2] **A**[3] **0**[4] 1. polymer 2. amount polymer
3.anti-corrosion 4. amount of anti-corrosion

To evaluate the corrosion resistance of each post-tensioning method, the experiment apparatus was designed as shown in Figure 1. The apparatus was slightly modified from the accelerated corrosion test (ACT) method specified in the Appendix B of PTI[7]. The chloride ion (Cl^-) concentration in the corrosion cell was 5%, made by dissolving sodium chloride (NaCl) to distilled water. Two series of experiments were conducted with the type of corrosion activation; 1) only chloride ion, and 2) both chloride ion and dissolved oxygen. In order to evaluate corrosion by chloride ion and dissolved oxygen, the concentration of dissolved oxygen was 21 ppm which was measured by dissolved oxygen meter (DO meter).

2.2 Electrochemical Corrosion Assessment

The corrosion of prestressing strand specimens protected by each corrosion protection method was electrochemically estimated, based on polarization resistance method. The polarization resistance, R_p, is defined as the ratio of the applied voltage, ΔE (shift in potential from E_{corr}), to the step of current, ΔI, when the metal is slightly polarized (about 20 ~ 50 mV) from its free corrosion potential, E_{corr}[8]. In this study, the potentiostatic method was used for corrosion measurement based on the polarization resistance method. A graphite rod was used as a counter electrode, and a saturated calomel electrode was used as a reference electrode. The working electrode was a prestressing strand specimen. The electrodes were connected to an electrostatic potential device, and the potential of the working electrode was measured as current applied for a short time period. The linear range of the current-potential curve is known to be linear in the range of 20 to 30 mV for the reinforcing steel. For the high corrosion rate, it is known to be linear in the range of 100 mV. Because there is a lack of research on the linear range of prestressing steel, a current corresponding to -10 ~ +10 mV, with respect to open-circuit potential, E_{oc}, was applied. The prestressing strand specimens were immersed in the corrosion solution for 60 days and the corrosion current density was measured every 15 days.

2.3 Weight Loss of Strands

To evaluate the degree of corrosion, the weight loss of prestressing strand due to corrosion was measured. The weight of strand was measured before and after the corrosion test. For the weighing, a precision balance with 0.1 mg of sensitivity was used to provide accurate measurement. After the ending of corrosion test, the prestressing strand in the specimen was taken and the impurities and corrosion products were removed. The prestressing strand was immersed in a cleaning solution and the corrosion product was removed using a resin brush. The cleaning solution was mixed with 20 g of antimony trioxide (Sb_2O_3) and 50 g of stannous chloride ($SnCl_2$) in 1 L of hydrochloric acid (HCl) solution. The cleaning procedure was referred to ASTM G1.

3. Experiment Results and Discussion

Figure 2 and Figure 3 represent corrosion current density of tendon protected by polymer, and anti-corrosion agent grouting with time, respectively. Grouting with polymer showed good performance to resist corrosion. This is proportional to the amount of polymer. For grouting with anti-corrosion agent, trend was not appeared. It should be noticed that the performance on corrosion had no relation with the amount of anti-corrosion agent.
Figure 4 shows the weight loss of the prestressing strand due to corrosion. The control specimens, which had no corrosion protection method, showed a very severe weight loss. The weight loss due to corrosion by chloride ion and dissolved oxygen was twice the weight loss by chloride ion solution. This has a good agreement with the result of corrosion current density. Likewise, the weight loss of the P7A0 P15A0 specimens showed a tendency similar to the measured corrosion current densities. Polymer can play a role of effective barrier to block the inflow of chloride ion or oxygen so that best corrosion protective performance was expected in the specimens. For the P0A0 specimens with no polymer and no anti-corrosion agent, the weight loss due to corrosion by chloride solution was similar to that of the P0A5 and P0A10 specimens. In a complex corrosive environment by combination of chloride ion and dissolved oxygen, however, the weight loss due to corrosion had increased significantly. It is concluded, therefore, that the corrosion protective performance of each corrosion protection method was consistent with the tendency found in the corrosion current density measurement. The corrosion resistivity was good in order of the P15A0, P7A0, P0A10, P0A5, P0A0, and the control.

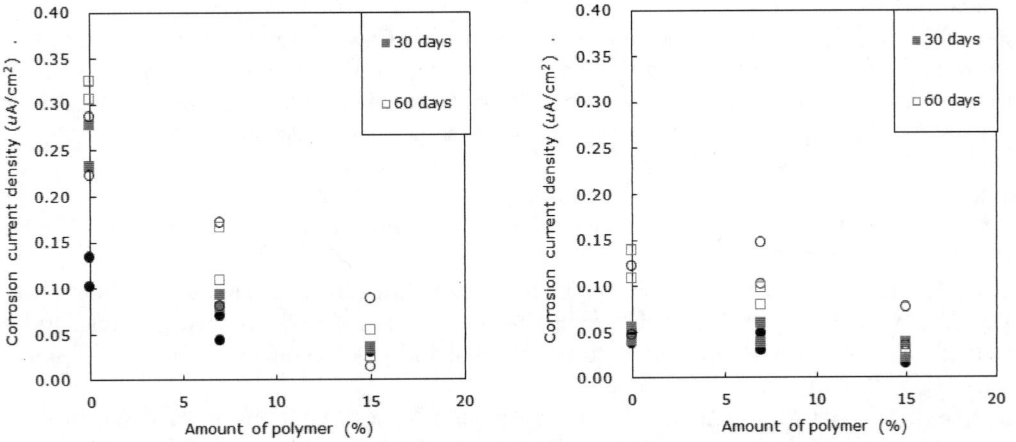

(a) Polymer-growing without oxygen (b) Polymer-growing with oxygen

Figure 2. Corrosion current density values according to the presence of oxygen (polymer)

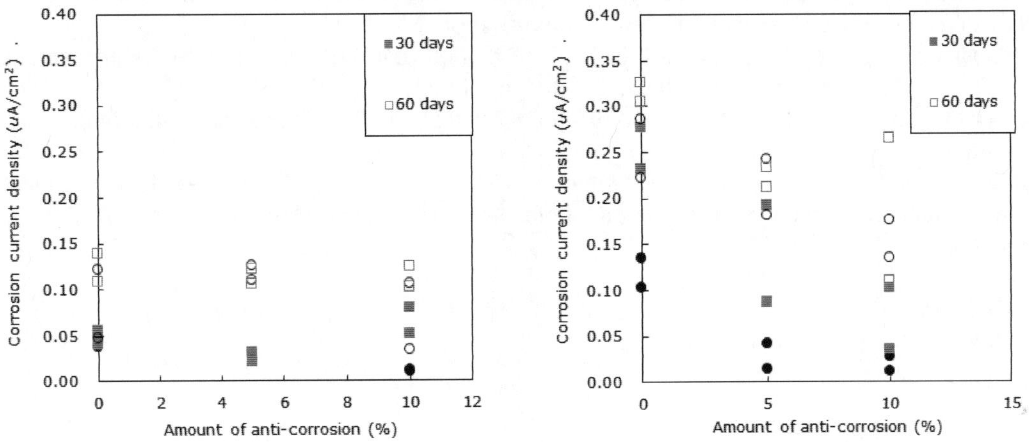

(a) Anti corrosion-growing without oxygen (b) Anti corrosion-growing with oxygen

Figure 3. Corrosion current density values according to the presence of oxygen (anti-corrosion)

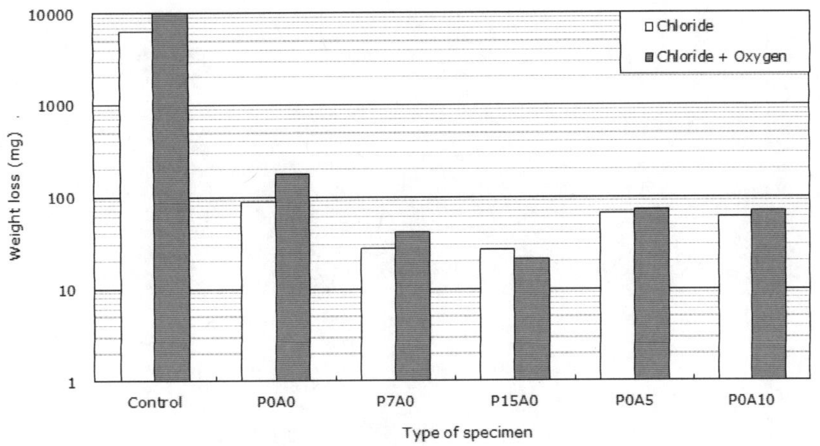

Figure 4. Weight loss of strands

4. Conclusions

The purpose of this study is to provide fundamental data for the reliability of corrosion protective systems. In comparison of the performance of each corrosion protective system based on the measurement of corrosion current densities and weight loss of prestressing strands, methods contained with ad the best performance for corrosion protection. The tendon protected by grouting with polymer was found superior to others.

5. References

[1] Powers, R. G. 1999. *Corrosion Evaluation of Post-Tensioned Tendons on the Niles Channel Bridge*, FDOT, Gainesville, FL.

[2] Trejo, D., Hueste, M. B., Gardoni, P., Pillai, R., Reinschmidt, K., Im, S. B., Kataria, S., Hurlebaus, S., Gamble, M., and Ngo 2009. T. T. Effect of Voids in Grouted, Post-Tensioned Concrete Bridge Construction: Vol.1-Electrochemical Testing and Reliability Assessment, Report No. 0-4588-1.TTI. College Station, TX.

[3] Hansen, B. 2007. Forensic Engineering: Tendon Failure Raises Questions about Grout in Post-Tensioned Bridges. *Civil Engineering News*, **77**(11), 17-18.

[4] Figg Bridge Engineers 2010. Inc., I-295. *Verina-Enon Bridge Tendon Replacement for the Virginia*, Department of Transportation, Exton.

[5] FHWA 2014. Post-Tensioning Tendon Grout Chloride Thresholds, An FHWA Special Study, US Dept. of Transportation Federal Highway Administration, McLean.

[6] O'Flaherty, F., Elomari I.R., Lambert, P. 2017. Corrosion Induced Losses in Prestressed Tendon, 37th Cement Concrete Science Conference, London, 111-115.

[7] PTI, PTI M55.1-12. 2012 Specification for Grouting of Post-Tensioned Structure. Post-Tensioning Institute, Farmington Hills.

[8] Andrade, C., and Alonso, C. 2004. Test Methods for On-Site Corrosion Rate Measurement of Steel Reinforcement in Concrete by Means of the Polarization Resistance Method. *Materials and Structures*, **37**(9) : 623-643.

Acknowledgments

This work was funded by KOFAC.

IOP Conf. Series: Materials Science and Engineering **559** (2019) 012003 doi:10.1088/1757-899X/559/1/012003

Functionalized Carbon-based Quantum Dots: Optical Characterization and Potential Application as Bio-fluorophore

R A T Cruz[1, 2,], A N Soriano[2], P A N de Yro[1, 2], G M O Quiachon[1], C S Emolaga[1], M L M Ysulat[1], U G Bigol[1] and B A Basilia[1, 2]

[1]Industrial Technology Development Institute, Department of Science and Technology, General Santos Avenue, Bicutan, Taguig City, 1631 Philippines
[2]School of Chemistry and Chemical Engineering, Mapua University, Intramuros, Manila, 1002 Philippines

E-mail:cruzrolandandrew@gmail.com

Abstract. Carbon quantum dots (CQDs), in comparison to heavy metal-based quantum dots offer renewable, non-toxic, low cost and easy synthesis production route while having excellent physicochemical properties for biomedical or environmental use. This paper discusses development and application of functionalized carbon quantum dots from glycerol as primary carbon source and tetraethylene pentamine as functionalizing agent. As-synthesized CQDs were characterized by Fourier Transform Infrared Spectrometer (FTIR), UV-Vis Spectrophotometer and Spectrofluorometer. FTIR spectra confirmed functionalization of resulting CQDs with emission wavelength peaks at 314.18 nm and 381.15 nm and observed strong blue luminescence under UV lamp. The performance of produced CQDs as bio-fluorophore for Gram-stained bacterial models was validated. Results indicated that CQDs can effect fluoresced images but a more distinct image can be observed on *S. aureus* compared to *E. coli*. An attachment mechanism of carbon quantum dots to the bacteria surfaces was also proposed here.

1. Introduction

The advent of the proliferation of fluorescent materials for various biomedical and environmental uses is leaning towards the adoption of naturally-based source materials. Currently, quantum dots, as these fluorescent materials are commonly called, are made up of organic dyes and heavy metals including cadmium selenide [1],[2]. Their uses in the biomedical and environmental field are restricted because of the hazard posed more than the benefits. Identification of carbon as possible alternative starting material for quantum dots is already being explored. Carbon quantum dots (CQDs) are nanosized materials measuring below 10 nanometers and are made up of quasi-spherical nanoparticles of amorphous to nanocrystalline core with hybridized carbon insertions. The CQDs apparent small size causes the variation of their properties as compared to their bulk material as described by the quantum effect mechanism. Smaller-sized CQD has higher band gap energy which means it requires higher amount of energy to excite the particle from its ground state due to stronger attractive forces. The CQD size must be small enough to effect quantum confinement thus forming surface emissive energy traps; and the addition of chemical functionalizing moieties provide stability to these energy traps that improve their properties [1]. These carbon quantum dots revolutionize the development of fluorescent

Content from this work may be used under the terms of the Creative Commons Attribution 3.0 licence. Any further distribution of this work must maintain attribution to the author(s) and the title of the work, journal citation and DOI.
Published under licence by IOP Publishing Ltd

materials by offering renewable, non-toxic, low cost and easy synthesis production route while maintaining excellent physicochemical & photochemical properties for intended biomedical or environmental uses. One potent application of CQD is its use as fluorophores especially in bacteriological studies. The CQD offers possible immediate detection or labelling and imaging of microorganisms present in various systems such as food or water supply due to their fluorescence under a fluorescent microscope [3].
This paper discusses the development of carbon quantum dots from glycerol as primary carbon source and tetraethylene pentamine (TEPA) as functionalizing agent. Also, the chemical make-up and optical characteristics of produced CQDs are determined and described in this study. Validation study for the application of the resulting CQDs as fluorophores for Gram-stained bacteria and their proposed attachment mechanism are reported here.

2. Methodology

2.1. Materials and apparatus

Glycerol (98.5% purity) and Tetraethylenepentamine (TEPA) were obtained from Sigma Aldrich. Both raw material chemicals were of analytical grade and were used as received without any further treatments or purification. A Tuttnauer Model 2340E autoclave with 200 L stainless steel chambers was also used for the prepared solutions of carbon quantum dots.

2.2. Synthesis and functionalization of carbon quantum dots

Carbon quantum dots were synthesized from glycerol through hydrothermal carbonization followed by chemical functionalization of the produced carbon core effected by TEPA. For the CQD solutions, a 15 ml glycerol was combined with 5 ml of TEPA or 25% TEPA in glycerol and; another 17.5 ml glycerol with 2.5 ml of TEPA or 12.5% TEPA in glycerol. Both glycerol-TEPA solutions were stirred for 15 minutes using magnetic stirrer to ensure mixture uniformity. These solutions were then transferred to a Teflon-lined stainless steel autoclave under the following operating conditions: temperatures of 180°C and 220 °C and; durations of 4 hours and 12 hours. The autoclaves were then cooled to room temperature. The resulting yellowish colored viscous solutions were collected and stored in glass sample bottles for analyses and application.

2.3. Optical characterization of synthesized cqds and performance testing as bio-fluorophore

The presence of specific functional groups in the chemical make-up of the produced CQDs were validated by a Fourier Transform Infrared Spectroscope (FTIR) Perkin Elmer Spectrum RX I over the wavenumber ranging from 400-4,000 cm-1 corresponding to the shifting of functional group peak readings. The maximum absorption wavelengths which correspond to the excitation spectra for the synthesized CQDs were determined using Ultraviolet-Visible Spectrophotometer (UV-Vis) UH5300. The emission wavelength of the produced CQDs was measured by Fluorescence Spectrophotometer Hitachi F-7000 which corresponds to the fluorescence measurement of each CQD solution. The E. coli and S. aureus bacterial strains were observed under Olympus BX51 microscope with fluorescence attachment using DAPI filter cubes of 350-370 nm excitation wavelength.The application of CQD as a bio-fluorophore was studied using Escherichia coli and Staphylococcus aureus as representative Gram-negative and Gram-positive bacterial model, respectively. These bacteria were subjected to in vivo staining technique and were observed under a fluorescence microscope for the image contrast effected by CQD to the model organisms.

3. Results and discussion

3.1. Characterization of CQDs

The FTIR spectra of the developed CQDs showed an observable broad band between 3000 to 3500 cm^{-1}. This indicated the presence of O-H bond with peaks centering at 3291.50 cm^{-1} and 3264.54 cm^{-1} which also suggested existent hydrogen bonding and vibrational stretching. The bending vibration of N-H bonds is observable at 1650 to 1580 cm^{-1} suggesting the conjugation of amine groups from TEPA.

Also, the presence of N-H bending in this region supports that N-H bonds exist for primary amine groups but were not observable due to weaker registers compared to broad O-H stretching which it overlaps with at 3300 to 3000 cm^{-1}. In the C-H band region at 2850 to 3000 cm^{-1}, C-H stretching for both developed CQDs is attributed to sp^3 hybridization of the carbon atoms. The presence of *sp* hybridized C-H bonds confirms that the stability of these nanoparticles can be attributed to the formation of sp^3 hybridizations in their molecular orbitals [2].

Table 1. FTIR spectra and corresponding functional groups

Wavelength	Functional Group
3000-3500 cm^{-1}	O-H stretching (Hydroxyls) N-H stretching (1° Amines due to N-H bending)
2850-3000 cm^{-1}	sp^3 hybridized C-H bond
1580-1650 cm^{-1}	N-H bending

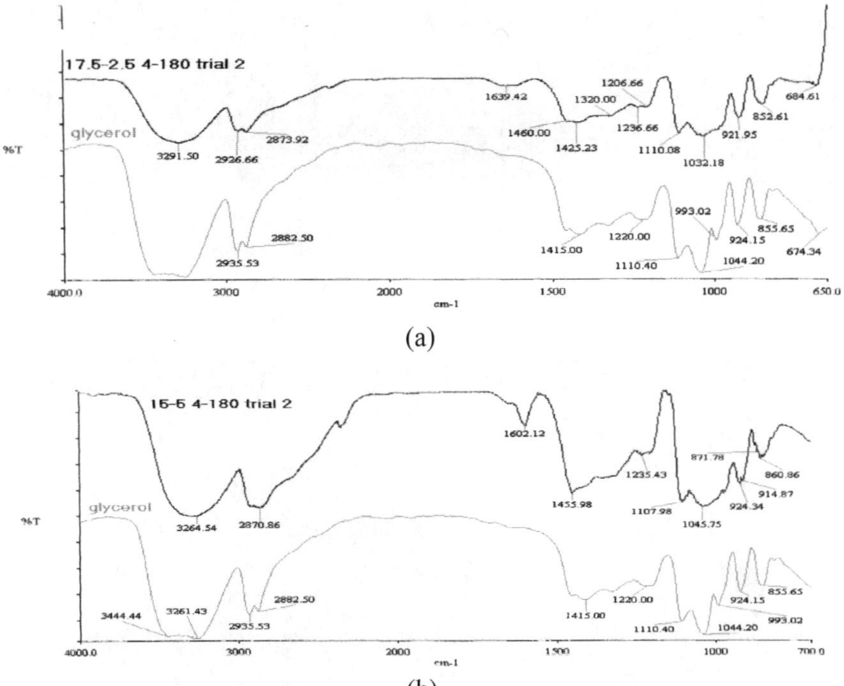

(a)

(b)

Figure 1. FTIR spectra of CQDs in the following conditions: (a) 12.5% TEPA in glycerol under 180°C for 4 hours and (b) 25% TEPA in glycerol under 180°C for 4 hours

The UV Vis spectra of developed CQDs recorded close maximum absorbance peaks of 207 and 210 nm, both wavelengths are under the UV region. This strong absorbance peak range in the UV region is suggestive of the π-π^* transition of conjugated π-bonds of CQDs. It was observed that some CQD solutions showed more than a single absorbance peak but were still within the identified UV absorption band.

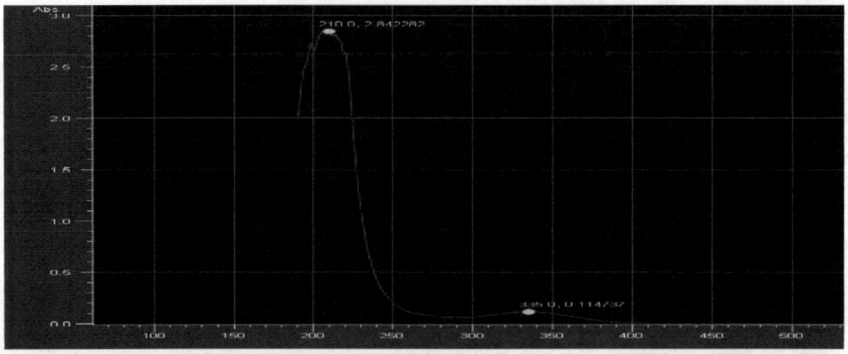

(a)

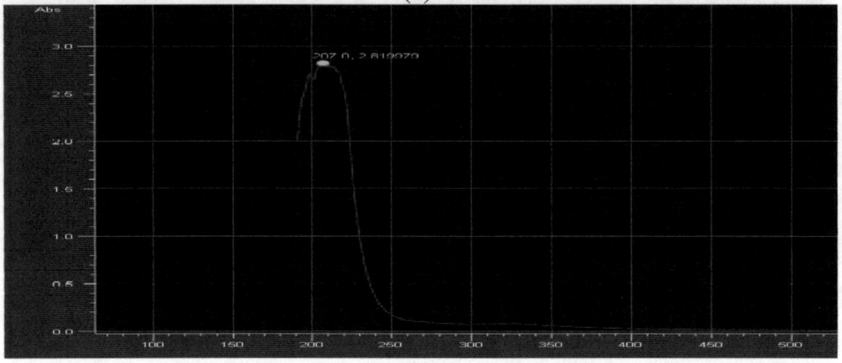

(b)

Figure 2. UV Vis spectra of CQD solutions prepared under the following process conditions: (a) 12.5% TEPA in glycerol under 220 °C for 4 hours and (b) 25% TEPA in glycerol under 220 °C for 12 hours

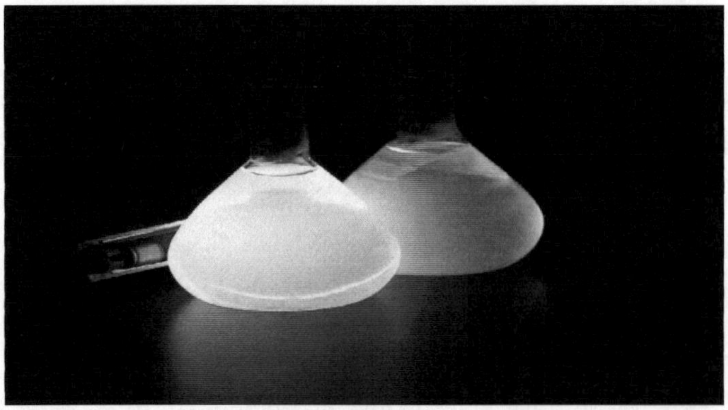

Figure 3. Developed CQD solutions showing strong blue luminescence under UV lamp

The developed CQDs were bombarded with an excitation wavelength of 210 nm taken from the maximum measured reading by the UV Vis Spectrophotometer. Maximum photoluminescence (PL) emission wavelengths peaked at 314.18 nm and 381.15 nm validating the strong blue luminescence under UV lamp with wavelength at 365 nm. The ability of CQDs to exhibit a strong blue fluorescence when placed under a UV lamp is also an indication of the formation of small-sized CQDs. Smaller CQDs have higher band gap energy which means it requires higher amount of energy to excite the particle from its ground state due to stronger attractive forces [4],[5]. This phenomenon results to shorter wavelength, higher frequency and closer luminescence to the blue end of the light spectra as observed in the developed CQDs in this study.

IOP Conf. Series: Materials Science and Engineering **559** (2019) 012003 doi:10.1088/1757-899X/559/1/012003

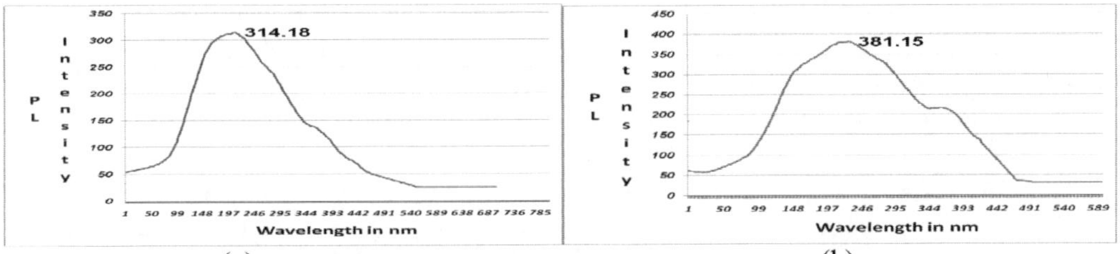

(a) (b)

Figure 4. PL Intensity of CQDs prepared under the following process conditions: (a) 25% TEPA in glycerol under 180 °C for 4 hours and (b) 12.5% TEPA in glycerol under 180 °C for 12 hours

3.2. Application as bio-fluorophore and plausible attachment mechanism

It can be seen in the fluorescence images of the bacterial models that CQDs effected a more observable fluoresced imaging of *S. aureus* compared to *E. coli*. This phenomenon can be attributed to the surface structures present in both microorganisms which allows for the unique interaction between CQDs and the Gram-stained bacteria suggestive of a plausible attachment mechanism.

(a) (b)

Figure 5. Fluorescence microscope images of representative Gram-stained bacteria with CQD solutions as imagers: (a) *Escherchia coli* (b) *Staphylococcus aureus*

Typical Gram-negative bacteria, such as *E. coli*, consist of an outer membrane made up of lipoproteins and lipopolysaccharides (i.e. phospholipids with negative polar head) over a thin, peptidoglycan wall. For Gram-positive bacteria like *S. aureus,* a singular but thicker peptidoglycan wall also exists, with observably absent phospholipid layer, interspersed with teichoic acid, which is an anionic polymer, on its surface. It was established that both Gram-stained bacteria have net negative charges based on these structures [6]. This strong blanket of negative charges enveloping *E. coli* and *S. aureus* attracts the positive particles of CQDs inducing temporary polarity that is supplemented by the formation of hydrogen bonds. A closer look to the chemical groups of the outer membranes of these bacteria show an abundance of amines (–NH) and hydroxyl (–OH) molecules common in amino- and carbohydrate-based structures. Similarly, chemical moieties of CQDs, also confirmed by the FTIR characterization, are mostly composed of hydroxyl (–OH) branching from carboxyl groups and amine (–NH) functional groups. These molecules tend to have greater negative charge density towards the nitrogen and oxygen ends due to high electronegativity and corresponding lone pairs. This behavior leaves the hydrogen atom with a positive charge density. The localization of positive charge in hydrogen atoms and deposition of negative charges on the molecules' electronegative end allows for the hydrogen to bond with the molecules' negative end. This intermolecular phenomenon enables CQDs to latch on to the bacterial surface thus effecting the fluoresced images. However, the complexity of the structure of *E. coli* due to the phospholipid-peptidoglycan bilayer compared to the single peptidoglycan layer of *S. aureus* offered an added resistance to the penetration of CQDs to the surface.

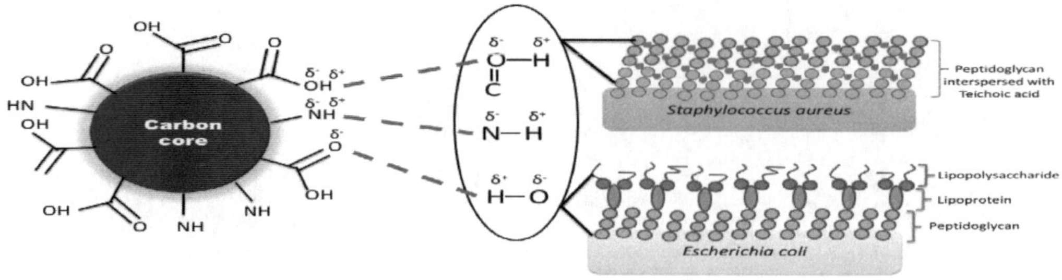

Figure 6. H-bonding as plausible attachment mechanism of CQD to Gram-stained bacteria

4. Conclusion

In conclusion, this study has successfully developed naturally-based quantum dots from glycerol as carbon source. The conjugation of amine groups sourced from TEPA to the CQDs was confirmed by FTIR spectra with the appearance of N-H bending at 1650 to 1580 cm^{-1}. The maximum excitation or absorbance wavelengths, recorded at 207 and 210 nm, which fall within the UV light range, are suggestive of the π-π^* transition of conjugated π-bonds of CQDs. Emission wavelengths peaked at 314.18 nm and 381.15 nm supporting the observed strong blue luminescence under UV lamp which is also an indication of small-sized CQDs formation. As a potential bio-fluorophore, developed CQDs effected more observable fluoresced image of *S. aureus* than that of *E. coli*. This phenomenon was attributed to the complex outer structure of *E. coli* which offered added resistance to the penetration of CQDs via hydrogen bonding as attachment mechanism. For future works, it is intended that the effect of incubation period of CQDs to bacterial models be assessed and the processing conditions be optimized to improve further CQD application as fluorophore.

5. References

[1] Liu X, Pang J, Xu F and Zhang X 2016 Simple approach to synthesize amino-functionalized carbon dots by carbonization of chitosan *Scientific Reports* **6** 31100
[2] Lv P, Yao Y, Zhou H, Zhang J, Pang Z, Ao K, Cai Y and Wei Q 2017 Synthesisof novel nitrogen-doped carbon dots for highly selective detection of iron ion *Nanotechnology* **28** (16): 165502
[3] Wang J, Zhang P, Huang C, Liu G and Leung K 2015 High Performance Photoluminescent Carbon Dots for In Vitro and In Vivo Bioimaging: Effect of Nitrogen Doping Ratios *Langmuir* ***31*** 8063-8073
[4] Abramowitz S 2017 *Handbook of Biosensors and Biochips* (Maryland, USA: John Wiley & Sons, Ltd.) p 133
[5] Fontes A, De Lira R, Seabra-Lopes M, Da Silva T, De Castro-Neto A and Santos B. 2012 Quantum Dots in Biomedical Research *Biomedical Engineering-Technical Applications Medicine* 269-289
[6] Nithish U and Sunitha S 2012 Bacteriological Applications of Quantum Dots *IJCRR* **4** 84-93

IOP Conf. Series: Materials Science and Engineering **559** (2019) 012004 doi:10.1088/1757-899X/559/1/012004

Design and Fabrication of NiTi Frame Utilizing for Endobronchial Valve

Supakit Amornthitipong[1], Thitiwat Sriprasart[2], and Anak Khantachawana[3, a]

[1] Biological Engineering Program, Faculty of Engineering, King Mongkut's University of Technology Thonburi (KMUTT), 126 Pracha Uthid Rd., Bangmod, Thung Khru, Bangkok, Thailand

[2] Interventional pulmonary, King Chulalongkorn Memorial Hospital, 1873 Rama 4 Rd., Pratumwan, Bangkok, Thailand

[3] Biological Engineering Program, Department of Mechanical Engineering, King Mongkut's University of Technology Thonburi (KMUTT), 126 Pracha Uthid Rd., Bangmod, Thung Khru, Bangkok, Thailand

E-mail: [a] anak.kha@kmutt.ac.th

Abstract. This study aims to design and fabricate the NiTi frame for Endobronchial valve (EBV) utilizing shape memory alloy wires. In order to evaluate the transformation and mechanical properties, heat-treatment at 450°C, 500°C and 550°C respectively for 10 minutes are carried out. Results from differential scanning calorimeter (DSC) and load-unloading test by universal tensile testing machine were used to select the heat-treatment condition for NiTi frame fabrication. Moreover, NiTi frames were fabricated by braiding 12 wires together by braiding machine with take up speed of 10, 15, 20 and 25 mm/sec respectively and constant carrier speed at 25 rad/sec. It is confirmed that all specimens show superior superelasticity at body temperature. Radial force of frame was measured by Mylar test at 37°C. Heat-treatment temperature at 500°C was selected since the highest stress for induce Martensite and recovery stress was obtained. The highest radial force was obtained in NiTi frame fabricated with taking up speed of 15 mm/sec. It is able to be concluded that NiTi frame can be illustrated for Endobronchial valve frame in future work.

1. Introduction

Chronic Obstructive Pulmonary Disease (COPD) is currently the fourth leading cause of death in the world and projecting to be the 3rd leading cause of death by 2020 [1]. COPD is a lung disease which is mainly caused by smoking and long-term exposure to air pollution, dust, and chemical fumes. It causes the airways in your lungs to inflame and thicken. The chronic airflow limitation that is characteristic of COPD is caused by chronic bronchitis and emphysema. Chronic inflammation makes the airways structural changes by narrowing the airways and destroying the lung parenchyma, that leads to loss of elastic recoil of alveolar. Moreover, these changes decrease the ability of the airways to exhale and remain open during expiration [2]. Emphysema also the main cause of excessive breathing difficulty. The exercise also decreased. In case of severity of the disease increase, even basic daily activity can make the patient feel tired all the time [3].

Nowadays Emphysema can be treated with many methods such as using Lung Volume Reduction Surgery (LVRS), etc. It is known that LVRS is one of the potential ways to treat emphysema, but the process of surgery requires a long time. On the other hand, Endobronchial valve (EBV) is another solution for Emphysema treatment, which is a minimally invasive treatment option for emphysema [4].

Content from this work may be used under the terms of the Creative Commons Attribution 3.0 licence. Any further distribution of this work must maintain attribution to the author(s) and the title of the work, journal citation and DOI.
Published under licence by IOP Publishing Ltd

EBV consists of NiTi frame which is made from Superelastic NiTi shape memory alloy (Nitinol) wires. Inside NiTi frame is attached with duckbill one-way valve to control airflow out of damaged lobe and block airflow into damaged lobe [5]. As a result of implanted EBV, the lung can reduce in volume and make damaged lung tissue can be healthier.
From above mentioned, since EBV is a new innovation for emphysema treatment and there is no publication report yet on the design and fabrication of EBV, it is a great challenge to study on design and fabrication of EBV, especially the locally made one which can be utilized physiologically for Asian people. This study focuses on the initial study of the effect of heat-treatment temperatures and braiding conditions on mechanical properties in order to fabricate superelastic EBV frame.

2. Materials and methods

NiTi with Ni-content of 50.8at% wires with diameters of 0.2 mm were used in this study as a raw material. Heat-treatment are carried out at temperature of 450°C, 500°C and 550°C for 10 min, respectively in vacuum furnace followed by rapidly quenching into cold water. DSC was conducted in order to obtain transformation temperature of specimen with different heat-treatment temperature. Load-unloading test was carried out with 'Mecmesin MultiTest 2.5i' testing machine with a 100 N load cell under temperature of human body temperature (37°C). Moreover, The NiTi frame was fabricated by braiding 12 NiTi wires together and forming into a tube shape with two difference cross-section, according to the human respiratory anatomy as shown in Table 1 [6].

Table 1. Cross-sectional area of airways.

Generation	Diameter (cm)
Trachea	1.80
Bronchi	1.22
Bronchioles	0.83
Terminal bronchioles	0.06

The braiding conditions for fabricating EBV frame are shown in Table 2. After braiding process, NiTi frame was constrained shape by heat-treatment process. Moreover, radial force of EBV frame was evaluated by Mylar film test. Figure 1 illustrates the fabricated EBV frame.

Table 2. Braiding conditions for fabrication EBV frame.

Wire numbers	Carrier speed (rad/sec)	Take up speed (mm/sec)
12	25	10
12	25	15
12	25	20
12	25	25

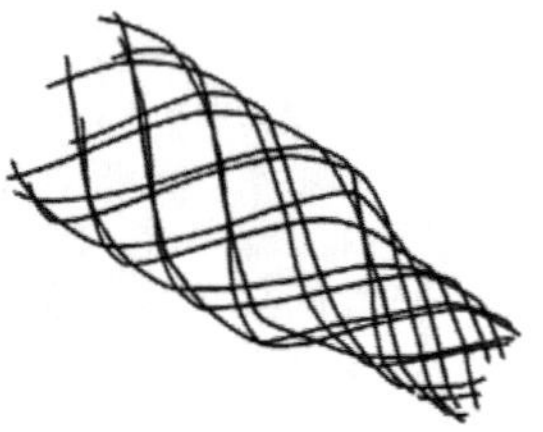

Figure 1. EBV frame from braiding process.

3. Results and discussion

Figure 2 reveals cooling and heating curves under constant cooling and heating rate of all specimens heat-treated at 3 difference temperature during DSC measurement. Two step transformation between Austenite phase, R-phase (Rhombohedral phase) and Martensite phase (Monoclinic phase) were clearly confirmed in specimens heat-treated at 450°C and 500°C while specimen heat-treated at 550°C shows the single step transformation between Austenite and Martensite phase. This R-phase formation can be explained by the residual stress during cold-work process in material manufacturing. R-phase disappears with increasing heat-treatment temperature above recrystallization temperature which is approximately 526°C. Although A_f temperature is not shown here, it is confirmed that A_f temperature of all specimens are below than 37°C or human body temperature which means that superelasticity can be confirmed.

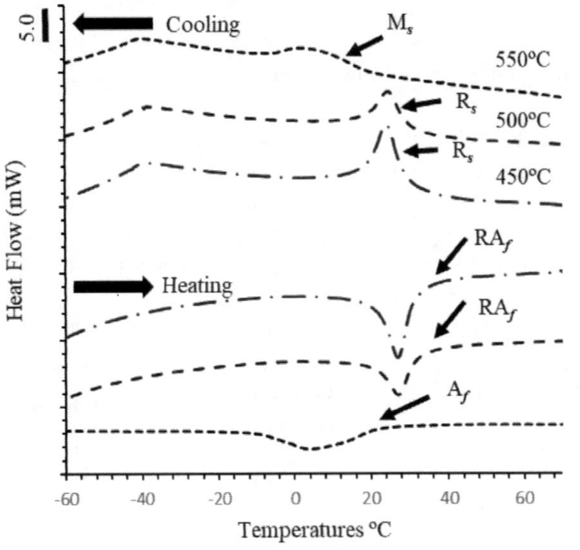

Figure 2. Cooling and heating curve of DSC measurement from specimens with difference heat-treatment temperature.

Table 3. Transformation temperature of specimens with difference heat-treatment temperature.

Heat treatment temperature (°C)	Transformation temperature(°C)			
	R_s	RA_f	M_s	A_f
450	29.60	16.5	-	-
500	30	17	-	-
550	-	-	-6	19

Figure 3 shows the stress obtained from load-unloading test of specimen with difference heat-treatment temperature. The upper and lower plateau stress are plotted against heat-treatment temperature. It is seen that upper plateau stress reveals maximum value at heat treatment of 500°C while lower plateau stress slightly decreases with increasing of heat-treatment temperature. From this result, heat-treatment at 500°C is selected to use as frame fabrication temperature since maximum stress can support the structure when the mass is similar.

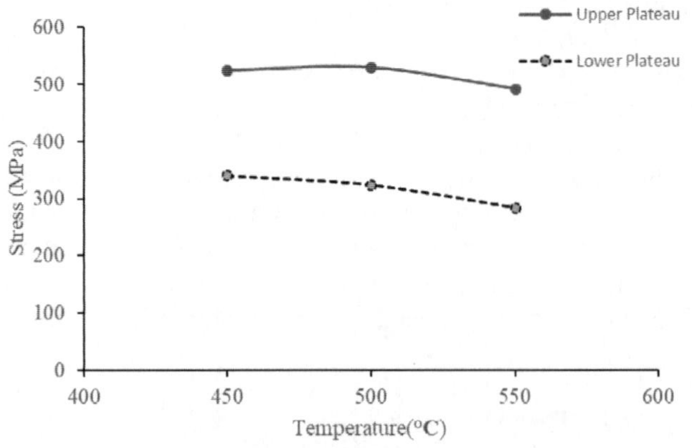

Figure 3. Upper and lower plateau stress of each specimen.

EBV frame was fabricated by braiding machine under various take up speeds, in the other word, various braiding angles. The braiding angle is observed with an optical microscope. Table 4 shows the relationship between take up speed and braiding angle. It is seen that braiding angle decreases with increasing take up speed. It is noted that NiTi frame can be constructed only when take up speed are higher than 10 mm/sec and lower than 25 mm/sec. Otherwise, shape will not be able to form.

Table 4. Braiding conditions with difference braiding angle.

Take up speed (mm/sec)	Braiding angle (Degree°)	Remarks
10	None	Cannot fabricate
15	45	
20	36	
25	26	Cannot fabricate

The Radial force of NiTi frame is performed by Mylar film test in order to obtain Radial Resistive Force (RRF) and Chronic Outward Force (COF) [7]. It is very important to evaluate the performance of NiTi frame under compression force from bronchial while implanted, also performance to return to the original shape of NiTi frame. Figure 4 shows the radial force obtained against braiding condition. It is obviously seen that RRF and COF significantly decrease with increasing of take up speed. Based on this data, the radial force of take up speed of 15 mm/sec will be designed and developed.

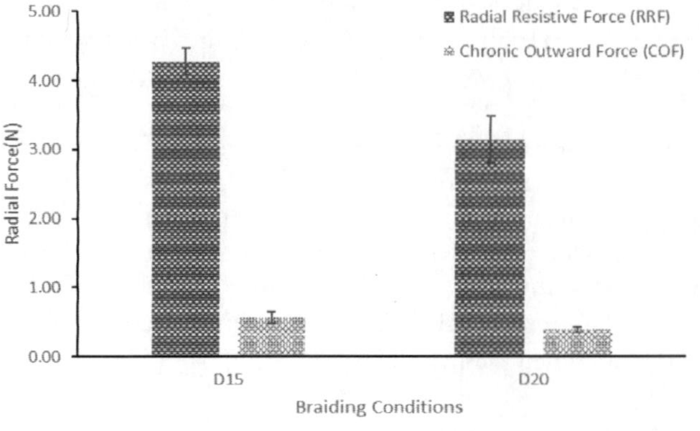

Figure 4. RRF and COF obtained from Mylar test of EBV frame with different braiding condition

4. Conclusion

In order to design and fabricate EBV frame from NiTi shape memory alloy wire, heat-treatment and braiding condition were investigated. It is able to summarize that heat-treatment at 500°C for 10 minutes should be the optimum condition since the maximum stress was obtained. Moreover, with increasing of take up speed braiding angle and radial force of NiTi frame decrease. The optimal take up speed for fabricating NiTi frame is 15 mm/sec with equivalent to 45° braiding angle. The further study on optimal force design of EBV device will be required.

5. References

[1] Vogelmeier C, Criner G, Martinez F, Anzueto A, Barnes P and Bourbeau J 2017 *Global strategy for the diagnosis, management, and prevention of chronic obstructive lung disease 2017 report* **195** 6-10

[2] Pornsuriyasak P, Amornputtisathaporn N and Kawamatawong T 2016 Proximal and peripheral airway morphologic changes in elderly patients with asthma with fixed airflow obstruction and COPD *European Respiratory Journal* **48** 740

[3] Meyers B and Bryan F 2002 Complications of lung volume reduction surgery *Seminars in thoracic and cardiovascular surgery* **14** 399-402.

[4] Sciurba F and Frank C 2010 A randomized study of endobronchial valves for advanced emphysema with advanced heterogeneous emphysema *New England Journal of Medicine* **363** 3-22

[5] Malka R et al. 2017 Pop-Up MEMS One-way endobronchial valve for treatment of chronic obstructive pulmonary disease *Journal of Medical Devices* **11** 1-10

[6] McCorry LK 2008 *Pulmonary Physiology* (New York, McGraw-Hill)

[7] Matsumoto T et al. 2016 Radial force measurement of endovascular stents: Influence of stent design and diameter *Vascular* **24** 171-76.

Acknowledgments

This research was supported by Smart Materials Research Laboratory, KMUTT, Innovation gap funding and National Research Council of Thailand (NRCT).

Bagasse Sheets Reinforced with Nanofibrillated Celluloses

S Klayya[1], P Rattanawongkun[1], N Kunfong[1], N Yodsuwan[1,2], N Tawichai[1,2], U Intatha[1,2], and N Soykeabkaew[1,2*]

[1]School of Science, Mae Fah Luang University,333 M 1, Muang, Chiang Rai 57100, Thailand
[2]Center of Innovative Materials for Sustainability (iMatS), Mae Fah Luang University, 333 M 1 Muang, Chiang Rai 57100, Thailand

Email: nattakan@mfu.ac.th

Abstract. The nanofibrillated celluloses (NFC) from different sources (i.e. bacterial cellulose (BC), pineapple leaf (PA) and banana pseudostem (BA)) were prepared using microfluidization. TEM and XRD results revealed diverse characteristics of the NFCs from different sources. Then, 0.1wt% of the prepared NFCs were integrated into the bagasse (BG) paper sheets. SEM images showed the densest surface of BG/NFC-BA sheet which also exhibited the highest sheet rigidity. Furthermore, the tensile tests indicated that the BG sheet reinforced with NFC-BC, possessing the highest fiber aspect ratio (L/D of 336) and crystallinity (80%), offered the highest strength and toughness. All tensile properties of the BG sheets were impressively enhanced with very low content (0.1wt%) of NFC addition. This confirmed that NFC is a highly effective reinforcement and suitable for use in paper making industry. However, suitable sources of NFCs for a particular paper product or application should be considered in advance.

1. Introduction

Nowadays, bagasse (BG) is one of important raw materials in paper and pulp industries. Papers and other related products made from BG pulp however are weaker than those made from wood pulp because BG has shorter fibers, hence, lower inter-fibers bonding [1]. Nanofibrillated cellulose (NFC) recently gains a great attention to use as a reinforcing additive to improve the physical and mechanicals properties of paper because of its unique properties such as biodegradability, high aspect ratio, high strength and high stiffness [2]. NFC can be prepared from many sources including biomass as well as bacterial cellulose (BC). It normally has a fiber diameter around 5-50 nm with a few micrometers length. NFCs extracted from different sources show specific fibers characteristic that can contribute to an effective improvement of paper quality [3-5]. The aim of this study is to assess the effect of NFCs from different sources (i.e. BC, pineapple leaf (PA) and banana pseudostem (BA)) reinforced in the BG paper sheets. The characteristic, morphology, structure, and properties of the NFCs and sheets were investigated using TEM, XRD, SEM, and tensile testing.

2. Materials and Methods

2.1. Materials

Bacterial cellulose (BC) was cultivated by *Komagataeibacter nataicola* in Hestrin-Schramm (HS) medium for 10 days before harvesting. Nanglae pineapple leaves used in this study were obtained from the field in Chiang Rai, Thailand. Cultivated banana pseudostem was taken from Mae Fah Luang

Content from this work may be used under the terms of the Creative Commons Attribution 3.0 licence. Any further distribution of this work must maintain attribution to the author(s) and the title of the work, journal citation and DOI.
Published under licence by IOP Publishing Ltd

IOP Conf. Series: Materials Science and Engineering **559** (2019) 012005 doi:10.1088/1757-899X/559/1/012005

University Botanical Garden, Thailand. Sodium hydroxide (NaOH) was an AR grade from QRëC, New Zealand. Bleached bagasse sheet was supplied by Biodegradable Packaging for Environment Public Co., Ltd., Thailand.

2.2. Preparation of BG pulp and sheet

BG paper was torn into small pieces and soaked in tap water overnight. The soaked BG papers were defibrillated by using kitchen blender (House Worth, HW-BDC2PC) at speed of 15,000 rpm for 5 min to prepare the BG slurry. Then, it was screened on the two metal meshes no.18 and 200 with opening sieve sizes of 1 mm and 74 μm, respectively. The portion remained on mesh no.18 was rejected and the pulp on mesh no.200 was collected, then adjusted to 2% pulp consistency. Next, the BG sheet was prepared by using hot-pressing technique under pressure of 159 kPa at 105°C for 5 min.

2.3. Preparation of NFCs from BC, PA, and BA

The harvested BC was cut into small pieces and disintegrated into the suspension with 0.5% concentration. To prepare NFC-BC, the BC suspension was passed through a microfluidizer (M-110P, Microfluidics) equipped with Y-type chamber (size of 87 μm) at 25,000 psi for 10 cycles. PA and BA were cut into 10 cm and dried until constant weight. Then, they were ground into small pieces and treated with 18% NaOH at 98±2 °C for 30 min using the liquid: solid ratio of 10:1. The obtained pulps were washed with tap water to remove black liquor until cleaned and pH adjusted to 7. The pulps were then disintegrated by using a kitchen blender (House Worth model HW-BDC2PC) at 15,000 rpm for 1 min before screening by the metal meshes no. 18 and 200. To produce NFC-PA and NFC-BAs, the pulp suspension was diluted to 0.5% consistency and refined using a microfluidizer (M-110P, Microfluidics), passing through Y-type chamber (87 μm channel) under pressure of 25,000 psi for 20 cycles.

2.4. Preparation of BG/NFCs sheets

Each NFC slurry was added to BG slurry at loading content of 0.1wt% (based on dried weight of BG) and then mixed for 1 min by stirring. Next, the mixture was preformed between two metal meshes (no.400) and then partially dewatered. The preformed sheet was then hot-pressed for 5 min at 105°C under pressure of 159 kPa.

2.5. Characterization

2.5.1. Scanning Electron Microscopy (SEM). The morphology of BG fibers and sheet surfaces were examined using scanning electron microscopy (SEM, LEO/1450 VP) at an accelerating voltage of 10 kV. All samples were coated with carbon by sputtering before observed. The dimension of BG fibers was measured and averaged from 50 fibers.

2.5.2. Transmission Electron Microscopy (TEM). Each NFC suspension was diluted to 0.001% consistency and then drop on a copper grid with formvar film support. The samples were then dyed with 2% uranyl acetate (UA) for 5 min and kept overnight before examined by transmission electron microscope (TEM, Hitachi Model HT7700) with acceleration voltage of 80 kV. The NFCs dimension was measured and reported from the average of 10 fibers.

2.5.3. X-ray Diffraction (XRD). NFCs in sheet forms were irradiated by CuKα at 40 kV and 30 mA with a symmetric reflection geometry in the angular range of 5°-35° (2θ) using a scan speed of 2°/min. The crystallinity index (CrI) of specimens was calculated using this following equation:

$$\%CrI = \left[\frac{(I_{002} - I_{am})}{I_{002}}\right] \times 100 \quad (1)$$

where I_{002} is the intensity reflection $(2\theta = 22°)$ and I_{am} is the intensity of amorphous cellulose $(2\theta = 18°)$ [1].

2.5.4. Physical properties. Each sheet sample was cut into 10 mm × 50 mm and measured for its thickness, width and length by using a digital Vernier caliper. Then, the volume (cm^3) was calculated and the weight (g) the sample was recorded according to a 4-digit balance. The density of each sheet sample was calculated by dividing its weight over volume. The grammage of paper sheet was calculated from the ratio of mass (g) to area of specimen (m^2). Porosity of each sample was calculated using this following equation:

$$\%Porosity = \left(1 - \frac{density\,of\ sheet}{density\,of\ cellulose}\right) \times 100 \qquad (2)$$

where the density of cellulose is 1.53 g/cm^3 [4].

2.5.5. Mechanical properties. Tensile testing was performed by using Instron model 5566 equipped with 1 kN load cell. The gauge length of specimen was 30 mm with width of 10 mm. All specimens were pre-conditioned at 50% relative humidity and 25°C overnight before testing. The machine was set to preload at 1 N before starting the test and cross-head speed was set at 15 mm/min. The testing was evaluated in term of tensile index, breaking length, elongation and Young's modulus. The results were obtained from the average of 5 tested specimens.

3. Results and Discussion

3.1. Characteristics of BG fibers and NFCs

SEM image of the BG pulp fibers is shown in figure 1A. The BG fibers have average length (L), diameter (D), and aspect ratio (L/D) of 274.8±105.9 μm, 11.7±4.9 μm, and 29, respectively.

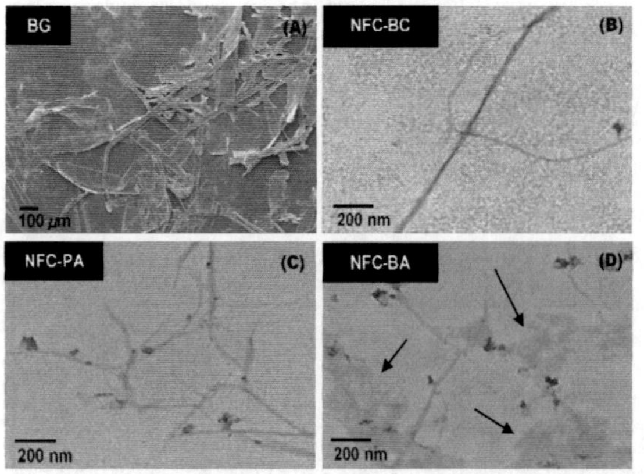

Figure 1. SEM photograph of BG fibers (A) and TEM photographs of the prepared NFCs (B-D).

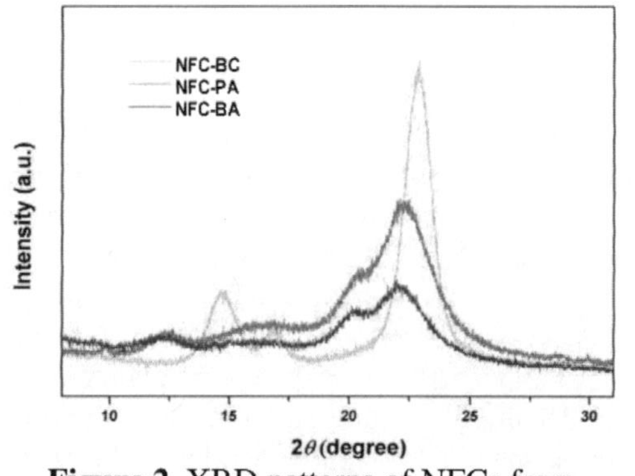

Figure 2. XRD patterns of NFCs from different sources.

Table 1. Dimension, aspect ratio and crystallinity (CrI) of NFCs from different sources.

Type of nanofibers	Length (μm)	Diameter (nm)	Aspect ratio	CrI (%)
NFC-BC	3.5±2.0	10.5±2.7	336	80
NFC-PA	1.7±0.5	15.5±3.6	108	69
NFC-BA	1.7±0.1	20.0±7.6	86	61

For NFCs, after 10 passes through the microfluidizer, BC was observed to be defibrillated into long single nanofibers (NFC-BC) (Figure 1B). On the other hand, after 20 passes, PA and BA microfibers were size-reduced to nanofibers of NFC-PA and NFC-BA, respectively (Figure 1C and 1D). The dimension and aspect ratio of the three NFCs are summarized in Table 1. It can be seen that NFC-BC showed the considerable higher aspect ratio than NFC-PA and NFC-BA. In addition, it seemed that NFC-BA nanofibers were partially damaged and some portions were turned into web-like networks (indicated with black arrows in Figure 1D).

XRD patterns of NFCs from BC, PA and BA were plotted together in figure 2. The sharp diffraction peaks at around 22.5° in all NFC samples designated the characteristic of cellulose I_β crystal structures at the (200) plane [6]. It was obvious that NFC-BC displayed the highest peak intensity and from the calculation using equation 1, this NFC had the highest crystallinity (CrI) about 80% (see Table 1). Perhaps, it was due to a difference in nature of cellulose sources. The lower CrI of NFC-PA and particularly NFC-BA might also indicate that after 20 passes through the microfluidizer, some portions of the crystalline nanofibers were partially damaged into amorphous phase which in line with the previous TEM results.

3.2. Morphology, Structure, and Properties of BG and BG/NFCs Sheets

With addition of only 0.1 wt% NFCs from different sources, the grammage of the obtained sheets was differently influenced and in the range of 240-340 g/m^2. Still, the density and porosity of the BG/NFC-BC and BG/NFC-PA sheets were quite similar to the pure BG sheet (see information in figure 3). Unlike the others, the BG/NFC-BA sheet showed a significant increase in density and decrease in porosity.

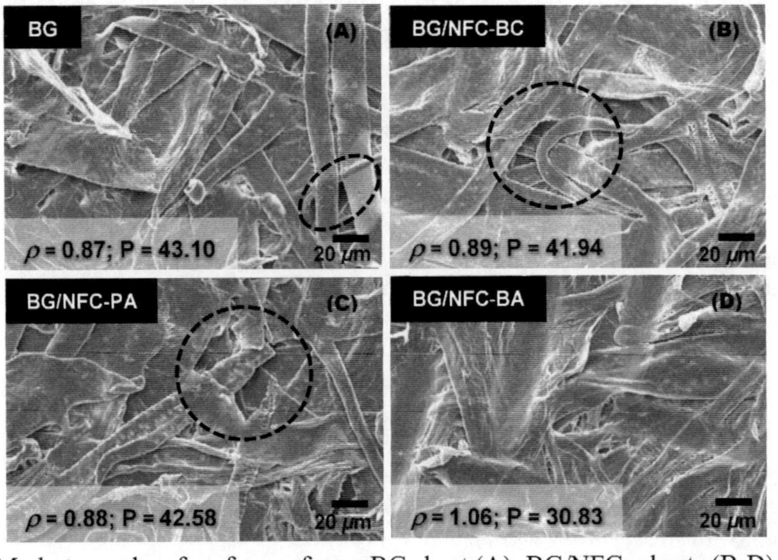

Figure 3. SEM photographs of surfaces of pure BG sheet (A); BG/NFCs sheets (B-D). Black circles indicate gaps between BG fibers. ρ = density (g/cm^3); P = porosity (%).

From the series of SEM images (Figure 3), the smoother surface of BG/NFC-BA when compared to other sheets can be observed. Possibly, the NFC-BA nanofibers with also web-like network parts (Figure 1D) can fill in more gaps between the BG microfibers, leading to less porous on surface and inside structure of this sheet.

The results from tensile tests are summarized in Table 2. Only 0.1wt% NFCs addition was able to improve all tensile properties of the BG sheets. The highest tensile index and breaking length was obtained with addition NFC-BC, then NFC-PA, and NFC-BA, respectively. Tensile index generally provides a measure of inter-fiber bonding within the paper [3]. It was likely that NFC-BC with the highest aspect ratio can bridge more fibers together and increase bonding ability of BG fibers in the sheet. The BG/NFC-BC also showed the highest elongation (ability of the fiber network to stretch under load until breaking). Elongation is also a good measure of the toughness of paper [7]. This enhancement was believed to mainly come from more areas of inter-fiber bonding and bridging within the sheet structure by NFC-BC [8].

Table 2. Tensile properties of pure BG sheet and BG/NFCs sheets.

Samples	Tensile index (Nm/g)	Breaking length (m)	Elongation (%)	Young's modulus (MPa)
BG	28.6 ± 5.3	2914.0 ± 535.1	1.6 ± 0.3	3060.2 ± 592.1
BG/NFC-BC	33.3 ± 3.9	3397.3 ± 400.8	2.8 ± 0.5	2764.4 ± 302.7
BG/NFC-PA	31.8 ± 6.5	3246.4 ± 663.2	2.1 ± 0.7	3102.0 ± 220.4
BG/NFC-BA	29.2 ± 2.8	2974.7 ± 282.0	1.9 ± 0.2	3421.0 ± 253.8

In case of Young's modulus which defines rigidity or stiffness of the sheets, the BG/NFC-BA sheet exhibited the highest value (Table 2). The densest structure of this sheet possibly created the strong initial bonding of fibers. Thus, at the beginning of the test, the sheet would be hardly elongated, leading to the high modulus value. However, the low aspect ratio of NFC-BA (Table 1) resulted in an early failure of this sheet.

4. Conclusion

NFCs from different sources (i.e. BC, PA, and BA) were successfully prepared by using a microfluidizer. Different sources of cellulose resulted in NFCs with diverse morphology, dimension, and structure (revealed by TEM and XRD results). The prepared NFCs were used to reinforce in BG paper sheets and it was found that only 0.1wt% NFC addition can improve all tensile properties of the resulting sheets. This improvement confirmed that NFC is a highly efficient reinforcing element for use in papermaking industry. The BG/NFC-BA sheet with the highest density and smoothest surface (observed by SEM) exhibited to be the stiffest (Young's modulus) sheet. On the other hand, the sheet integrated with NFC-BC, having the highest aspect ratio and crystallinity, was shown to have the highest strength (tensile index and breaking length) and toughness (elongation) followed by the BG/NFC-PA and BG/NFC-BA sheets, respectively. This can bring a conclusion that different NFCs characteristic clearly affected their reinforcing efficiency in the BG sheets. Therefore, a selection of suitable NFC sources for desired performance of papers and other related products has to be considered beforehand.

5. References

[1] Petroudy S R D, Syverud K 2014 *Carbohydr Polym.* **99** 311
[2] Gonzalez I, Vilaseca F 2013 *Cellulose.* **20** 1425
[3] Campano C, Merayo N 2018 *Int J Biol Macromol* **114** 1077
[4] Salas C, Nypelo T 2017 *J Colloid and Interf* **101** 383
[5] Xiang Z, Jin X 2017 *Cellulose* **24** 5147
[6] Xiaohui J, Mark B 2015 *Carbohydr Polym* **123** 476
[7] Zeng X, Vishtal A 2013 *BioResources* **8** 472
[8] Tabarsa T, Sheykhnazari S 2017 *Int J Biol Macromol* **101** 334

Acknowledgments

The joint funding support (grant code 256101A3070019) from the National Research Council of Thailand (NRCT) and National Science Technology and Innovation Policy Office (STI) is gratefully acknowledged. All supports and facilities from our industrial partner, the Biodegradable Packaging for Environment Public Co., Ltd. (Gracz) as well as Mae Fah Luang University (MFU) and the PLC Holding Co., Ltd., Thailand are also much appreciated.

IOP Conf. Series: Materials Science and Engineering **559** (2019) 012006 doi:10.1088/1757-899X/559/1/012006

Nanofibrillated Cellulose from Banana Pseudostem Reinforced Bagasse Paper Sheets

N Soykeabkaew[1,2], P Rattanawongkun[1], N Kunfong[1], N Tawichai[1,2], and U Intatha[1,2]

[1]School of Science, Mae Fah Luang University,333 M 1, Muang, Chiang Rai 57100, Thailand
[2]Center of Innovative Materials for Sustainability (iMatS), Mae Fah Luang University, 333 M 1, Muang, Chiang Rai 57100, Thailand

Email: nattakan@mfu.ac.th

Abstract. Underutilized agro-waste of cultivated banana pseudostem was treated by a soda pulping process and passed through a microfluidizer to prepare nanofibrillated cellulose (BA). TEM images confirmed the presence of thin fibrils and fine webs of BA with average diameter and length of 20.0±7.6 nm and 1.7±0.1 μm, respectively. Reinforcing potential of this BA nanofibers in recycled papers was investigated by adding 0.5, 1.0 and 5.0 wt% BA into the recycled bagasse sheets. Tensile testing results indicated that all mechanical properties (i.e. tensile index, breaking length, Young's modulus, and elongation) were noticeably increased all at once. SEM images revealed that the physically retained BA nanofibers in the sheet structures bridged and linked between bagasse fibers increasing inter-fiber bonding, hence, strengthening the sheets as a result. Toughness of the reinforced sheets was also found to be significantly enhanced (~128% increase with 5.0 wt% BA incorporation). The change in failure mode showing fibers stretching and slippage before breaking was observed on fractured surface of this paper sheet. This BA cellulose nanofiber demonstrated to be highly promising as reinforcement in pulp and paper industry.

1. Introduction

The recovery of paper and board is a well-established practice these days as a result of concerns about environmental issues, energy saving, population growth, and the shortage of wood supplies [1-2]. By increasing the number of recycling cycles, however, fibers experience structural damages due to mechanical refining and hornification phenomena which reduces paper quality, especially its strength [3]. Recently, an addition of cellulose nanofibers has emerged as a new approach to improve the strength of paper and board due to their attractive characters such as high specific surface, high aspect ratio, high intrinsic mechanical properties, renewable nature, biodegradability, high potential availability, and capacity to be functionalized [1]. The incorporation of cellulose nanofibers has been reported to significantly increase bond degree and paper strength as well as extend the life span of paper products to more than twice [3]. The main methods currently used to effectively produce cellulose nanofibers or nanofibrillated cellulose are the high pressure homogenization and the micofluidization [4]. Cellulose nanomaterials are usually extracted from wood, however, fibers from annual crops and agricultural residues has attracted more and more interest due to their wide abundance, low cost, and availability [5].

The main goal of this study is to examine the potential of nanofibrillated cellulose prepared from banana pseudostem (BA) to be used as a reinforcing material in recycled bagasse papers. Mechanical

Content from this work may be used under the terms of the Creative Commons Attribution 3.0 licence. Any further distribution of this work must maintain attribution to the author(s) and the title of the work, journal citation and DOI.
Published under licence by IOP Publishing Ltd

IOP Conf. Series: Materials Science and Engineering **559** (2019) 012006 doi:10.1088/1757-899X/559/1/012006

and surface properties as well as porosity of the reinforced paper sheets, which are interesting from an industrial point of view, were characterized.

2. Experimental

2.1. Materials

Cultivated banana pseudostem was taken from Mae Fah Luang University Botanical Garden, Thailand. Bleached bagasse sheet was supplied by Biodegradable Packaging for Environment Public Co., Ltd., Thailand. Sodium hydroxide (NaOH) was an AR grade from QRëC, New Zealand.

2.2. Preparation of nanofibrillated cellulose from banana pseudostem (BA)

Firstly, the banana pseudostem pulp was prepared by soda pulping process. The pseudostem was chopped into 10 cm length then washed and dried in a hot air oven until constant weight. Next, the dried pseudostem was size-reduced by using a pulverizer (Nimut Engineering Co., Thailand) and treated with 18% NaOH at 98±2 °C for 30 min using the liquid:solid ratio of 10:1. The obtained pulp was then rinsed by tap water to remove black liquor until cleaned and pH adjusted to 7. The pulp was disintegrated by using a kitchen blender (House Worth model HW-BDC2PC) at 15,000 rpm for 1 min before screening by the metal meshes no. 18 and 200 with opening sieve sizes of 1 mm and 74 μm, respectively. The collected pulp for further use was the screened portion between the two metal meshes. To prepare nanofibrillated cellulose, the previous pulp was diluted to 0.5 wt% and homogenized through the interaction chamber of 87 μm at 25,000 psi for 20 passes using an M-110P microfludizer® processor (Microfluidics, U.S.A.).

2.3. Preparation of bagasse pulp

The bagasse sheet was torn into small pieces (ca. 1 cm × 1 cm) and dispersed in tap water at 2% consistency overnight. Then, the bagasse suspension was disintegrated by using a kitchen blender (House Worth model HW-BDC2PC) at 15,000 rpm for 5 min before screening by the metal meshes no. 18 and 200. The screened pulp portion between the two metal meshes was collected and prepared into a suspension with 2% consistency for further use.

2.4. Preparation of bagasse and bagasse/BA paper sheets

The bagasse pulp slurry was directly added to the BA suspension in ratios of 0.1, 0.5, 1.0, and 5.0 wt% and gently stirred and mixed for 1 min. Then, each mixture was partially dewatered by hand press between the two metal meshes no. 400 and hot-pressed into a paper sheet at 105 °C for 5 min under 159 kPa pressure. The bagasse sheet without BA addition was also formed for comparison. The grammage of all paper sheets made were in the range of 220-280 g/m^2 having density and porosity in the ranges of 0.81-1.00 g/cm^3 and 34.4-47.1 %, respectively.

2.5. Bagasse pulp and BA characterization

The bagasse pulp fibers were characterized by scanning electron microscopy (SEM) (LEO 1450 VP, U.S.A.) at an accelerating voltage of 10 kV. Gold was sputtered on samples prior to the observation. The morphological analysis of the BA nanofibers was carried out using a transmission electron microscope (TEM) (Hitachi HT7700, Japan) at an accelerating voltage of 80 kV. For sample preparation, a BA suspension of 0.001 wt% was deposited to a copper grid with formvar film and the droplet was maintained on the grid for 5 min, and then strained with a 2% aqueous uranyl acetate.
The diameter and length of bagasse pulp and BA fibers were measured from SEM and TEM images, respectively. The reported values were averaged from 50 bagasse fibers and 10 BA fibers.

2.6. Paper sheet testing and characterization

Tensile testing of the paper sheets was carried out using Instron (model 5566) testing machine (U.S.A) equipped with a 1 kN load cell. The width and gauge length of specimens were 10 mm and 30 mm, respectively. The cross-head speed was set at 15 mm/min and the test started with a preloaded of 1 N. Before testing, all specimens were conditioned at 50% relative humidity (RH) at 25 °C overnight.

IOP Conf. Series: Materials Science and Engineering **559** (2019) 012006 doi:10.1088/1757-899X/559/1/012006

The surfaces and fractured surfaces (after the tensile test) of the paper sheets were characterized by using SEM (LEO 1450 VP, U.S.A.) at an accelerating voltage of 10 kV. A thin layer of gold was coated on sheet samples prior to the examination.

3. Results and Discussion

Figure 1 shows SEM (a) and TEM (b) images of bagasse pulp fibers and BA nanofibrillated cellulose fibers, respectively. The bagasse pulp fibers had an average diameter of about 11.7±4.9 μm and length of 274.8±105.9 μm with the aspect ratio around 29. On the other hand, after passing through the microfluidizer (20 passes), the BA nanofibers with an average diameter of 20.0±7.6 nm and length of 1.7±0.1 μm, having an aspect ratio around 86, were obtained. Some elementary fibrils (ca. 5 nm diameter) fibrillated from single BA nanofibrils were also observed (Figure 1b). Typically, single cellulose nanofibrils were reported to have a diameter of 5-50 nm and a length of few micrometers depending on processing conditions, sources, etc. [6].

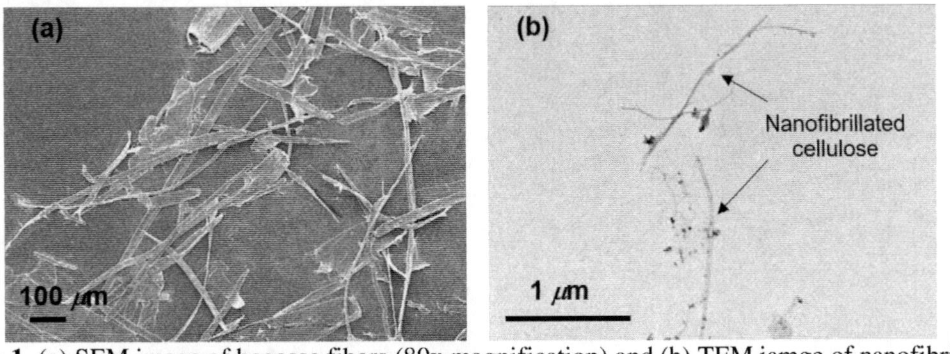

Figure 1. (a) SEM image of bagasse fibers (80x magnification) and (b) TEM iamge of nanofibrillated cellulose from banana pseudostem (8,000x magnification).

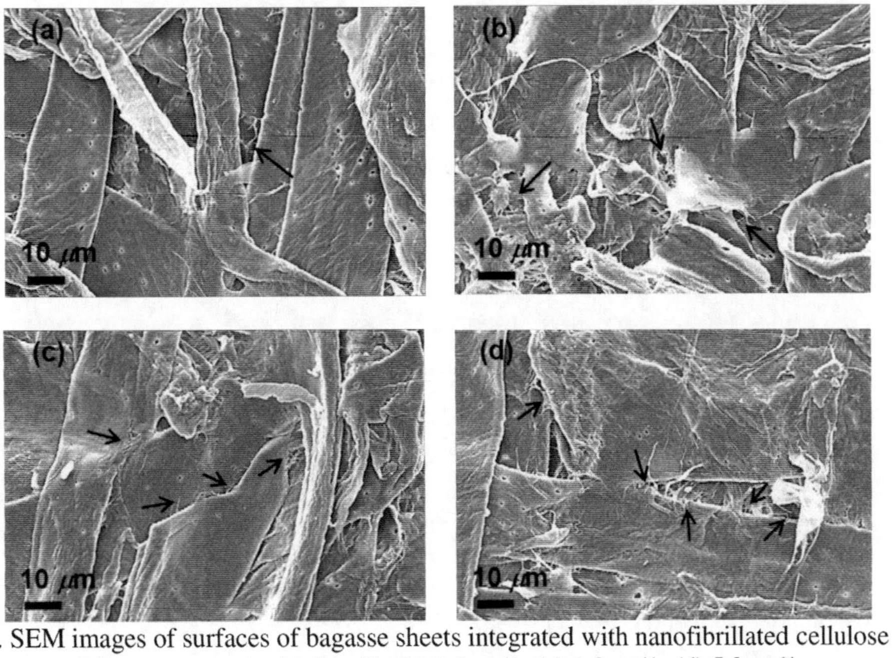

Figure 2. SEM images of surfaces of bagasse sheets integrated with nanofibrillated cellulose (BA) at different contents. (a) 0 wt%; (b) 0.5 wt%; (c) 1.0 wt%; (d) 5.0 wt%.

SEM was used to examine the surface morphology of the pure bagasse sheet and the BA reinforced paper sheets. Micrograph obtained from the pure sheet (Figure 2a) revealed the large gaps/voids between the fiber network. Few fines (thin fibrils) were also observed in this bagasse sheet as indicated with the black arrow. With increasing BA content, the presence of fines was raised confirming the retention of BA within the bagasse paper sheets (Figure 2b-d). These BA nanofibers were found to retain in the gaps between adjacent fibers, joining these fibers and also covered some parts of the fiber surfaces. Bridging of these fibers by BA creates a better bonding within the paper structures leading to a strengthening effect of the sheets [7,8]. This results are in good agreement with the increased mechanical properties of the reinforced paper sheets which are about to discuss in the following section.

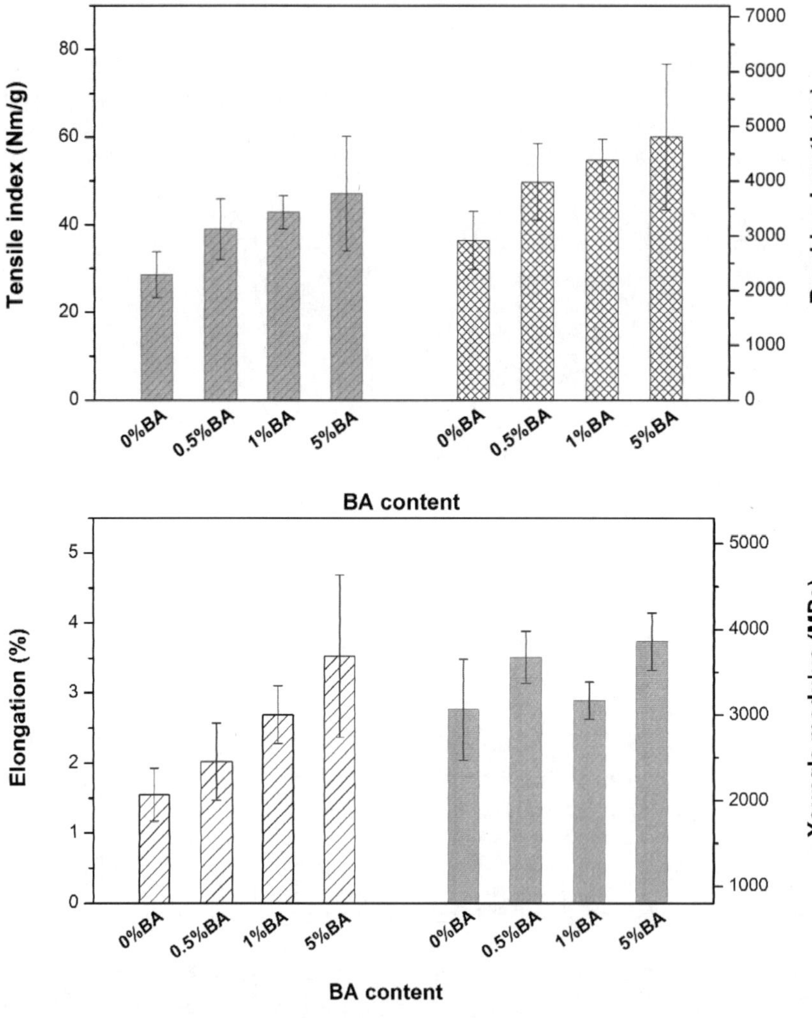

Figure 3. Mechanical properties of pure bagasse sheet (0%BA) and the sheets integrated with 0.5-5.0 wt% of nanofibrillated cellulose from banana pseudostem (0.5%BA-5%BA).

From the tensile testing results (Figure 3), the reinforcing effect of BA was confirmed by the progressively improvement in the tensile index as well as the breaking length when the BA content in the paper sheets increased. Only 1.0 wt% BA addition could significantly increase both the tensile index (to 43 Nm/g) and breaking length (to 4377 m) of the paper sheet by 50%. The similar result has been recently reported by Espinosa et al. [3]. In their previous work, the incorporation of 1.5 wt% lignocellulosic micro/nanofibers (LCMNF) from banana leaf residue was shown to enhance the breaking length of recycled papers from 2749 m to 3751 m (~36% increment) which became greater than the value requirement for using as fluting papers (> 3443 m).

Figure 4. Scanning electron micrographs of fractured surfaces of bagasse sheets integrated with (a) 0 wt% and (b) 5 wt% nanofibrillated cellulose (BA).

Elongation is defined as the ability of the fiber network to stretch under load until breaking and is a good measure of the toughness of paper [8]. With increasing the BA content, the elongation of the current paper sheets increased considerably. This enhancement was believed to mainly come from a higher areas of inter-fiber bonding and bridging within the sheet structures by BA nanofibers as revealed by the previous SEM results (Figure 2). From the fractured surfaces comparing between the pure bagasse sheet (Figure 4a) and the 5.0 wt% BA integrated sheet (Figure 4b), differences in breaking mode of these two specimens were clearly seen. The failure with low degree of fiber stretching was observed in the pure bagasse sheet while, in the failed sheet with BA addition, yielding of the fibers and fines as well as slippage among them was noticed. Additionally, an increasing trend of the Young's modulus (rigidity) of the sheets was also presented with higher BA contents (Figure 3).

4. Conclusions

Nanofibrillated cellulose from banana pseudostem (BA) was successfully prepared and used to reinforced in recycled bagasse paper sheets at 0.5, 1.0 and 5.0 wt%. SEM images revealed that BA nanofibers were physically retained on the sheet surfaces, bridging and connecting bagasse fibers together. This indicated an increase in fiber bonding areas within the sheet structures. All tensile properties of the BA reinforced sheets were noticeably enhanced, particularly their elongation or toughness. The BA integration led to a change in failure mode of the reinforced sheets allowing fibers to stretch and slide further before breaking. The current BA cellulose nanofiber demonstrated good potential as a reinforcing element for use in pulp and paper industry.

References

[1] Tarres Q, Pellicer N 2017 *Int J Biol Macromol.* **105** 664
[2] Bale A, Blanco A 2016 *BioResources.* **11** 8123
[3] Espinosa E, Tarres Q 2018 *J Clean Prod.* **17** 233
[4] Boufi S, Gonzales I 2016 *Carbohydr Polym.* **154** 151
[5] Petroudy S R D, Syverud K 2014 *BioResources.* **99** 311
[6] Nechyporchuk O, Belgacem M N 2016 *J. Bras: Ind Crops Prod.* **93** 2
[7] Campano C, Merayo N 2018 *Int J Biol Macromol* **114** 1077
[8] Tabarsa T, Sheykhnazari S 2017 *Int J Biol Macromol.* **101** 334

Acknowledgments

The funding support (code 256101A3070019) from the National Research Council of Thailand (NRCT) and National Science Technology and Innovation Policy Office (STI) is gratefully acknowledged. All supports and facilities from the Biodegradable Packaging for Environment Public Co., Ltd. (Gracz) as well as Mae Fah Luang University and the PLC Holding Co., Ltd., Thailand are also much appreciated.

IOP Conf. Series: Materials Science and Engineering **559** (2019) 012007 doi:10.1088/1757-899X/559/1/012007

Micro/Nano Papers from Bagasse Pulp Reinforced by Bacterial Cellulose Nanofibers

P Rattanawongkun[1], N Kunfong[1], N Tawichai[1,2], U Intatha[1,2], N Yodsuwan [1], and N Soykeabkaew[1,2]

[1]School of Science, Mae Fah Luang University, 333 M 1, Muang, Chiang Rai 57100, Thailand

[2]Center of Innovative Materials for Sustainability (iMatS), Mae Fah Luang University, 333 M 1, Muang, Chiang Rai 57100, Thailand

Email: nattakan@mfu.ac.th

Abstract. The aim of this research was to used bacterial cellulose (BC) as a biodegradable reinforcing agent in bagasse (BG) paper sheets. BC was cultured by *Komagataeibacter nataicola* and defibrillated into isolated nanofibers by using microfluidizer. Micro/Nano (BG/BC) papers were prepared by adding 0.5 and 5 wt% BC in BG pulp and then the paper sheets were fabricated by compression molding. The morphology, physical and tensile properties of the prepared sheets were studied. It was found that the adding BC increased the density of BG sheets while their porosity was decreased. SEM images confirmed that BC filled gaps and bridged between BG fibers leading to a raise in fiber bonding. With increasing BC content, the mechanical properties of paper sheets were clearly improved. The addition of 5 wt% BC significantly enhanced the tensile strength and elongation at break of the reinforced sheets by 47% and 117%, respectively. Furthermore, the tensile index and breaking length of the BG sheets were increased for approximately 35%.

1. Introduction

The production of specialty paper products such as high-strength papers requires an additive or strengthen binder to enhance their properties. Bacterial cellulose (BC) has been investigated as a paper binder because it has several benefits over plant cellulose such as high purity, high tensile strength and high crystallinity [1, 2]. In previous works, BC has been successfully used as a paper additive in bleached spruce sulfite pulp for the high-strength functionalized paper sheets as well as in softwood sheets [2, 3]. Besides wood, non-wood resources such as rice straw, bamboo and bagasse are the available fiber sources and has potential to produce suitable pulps for various applications [4]. Bagasse (BG), the fibrous residue that remains after sugarcane stalks are crushed to extract their juice, is one of the largest agriculture residues in the world and currently also used in the manufacture of pulp. This work aims to study the feasibility of using defibrillated BC as a reinforcement in the BG paper sheets. The Micro/Nano (BG/BC) papers were fabricated and characterized in term of their morphology, physical and mechanical properties.

2. Experimental

2.1. Preparation of bacterial cellulose nanofibers

Bacterial cellulose (BC) was cultivated by *Komagataeibacter nataicola* in the medium composed of sucrose 50 g, yeast extract 5 g, KH_2PO_4 3 g and $MgSO_4.7\ H_2O$ 0.05 g dissolved in 1 L of water. The

Content from this work may be used under the terms of the Creative Commons Attribution 3.0 licence. Any further distribution of this work must maintain attribution to the author(s) and the title of the work, journal citation and DOI.
Published under licence by IOP Publishing Ltd

produced BC was harvested to prepare BC suspension in water at a concentration of 0.5% for 400 mL and then homogenized using a microfluidizer (M-110P, Microfluidics) equipped with the Y-type chamber (size of 87 μm) at 25,000 psi for 10 cycles.

2.2. Preparation of bagasse pulp

Bagasse (BG) was obtained from Biodegradable Packaging for Environment Public Co., Ltd (BPE), Thailand. BG paper was torn into small pieces and soaked in tap water overnight. The soaked BG papers were defibrillated by using a kitchen blender (House Worth, HW-BDC2PC) at speed of 15,000 rpm for 5 min. The BG slurry was screened on the metal mesh sieves with 1 mm slits and 0.074 mm. The pulp remained on the 1 mm slit sieve was rejected. The screened pulp on the 0.074 mm slit sieve was collected and then adjusted to 2% pulp consistency.

2.3. Preparation of micro/nano (BG/BC) papers

To prepare the Micro/Nano paper sheets, the mixtures of BG and BC slurries were prepared. The BC was added into the BG slurry with the varied contents of 0, 0.5, 5 wt% and then mixed for 1 min by hand. The well-dispersed mixture was pre-formed into a wet sheet between two metal meshes (no.400) and water was then partially removed out. After that, the wet sheet was hot-pressed for 5 min at 105ºC under pressure of 159 kPa to obtain the dried paper sheet.

2.4. Characterization

2.4.1. Morphology. Scanning electron microscope (SEM) (LEO/1450 VP) was used to observe the characteristic of BG fibers and the morphology of prepared sheet surfaces with an accelerating potential of 10 kV. Before observation, BG fibers was diluted in water and adjusted into 0.2% pulp consistency, then dropped onto a transparent sheet and placed in a desiccator until dried. The dimension of BG fibers was measured and averaged from 50 fibers. For the paper sheets, they were cut into the size of 5 mm × 5 mm and coated with carbon by sputtering before examination. On the other hand, the morphological analysis of BC was carried out using a transmission electron microscope (TEM) (Hitachi Model HT7700) with an acceleration voltage of 80 kV. The BC suspension was diluted to 0.001% consistency and dropped on a copper grid with formvar film support. All samples were then dyed with 2% uranyl acetate (UA) for 5 min and kept overnight before observation. The dimension of BC was measured and averaged from 10 fibers.

2.4.2. Physical properties. To assess the density of the paper sheets, they were cut into rectangular specimens with the dimension of 10 mm × 50 mm. Then the thickness, width and length of each specimen were carefully measured by using a digital Vernier caliper. Next, the volume *(*cm^3*)* of each specimen was calculated from its measured dimension. Also, the weight *(g)* of each specimen was recorded by a 4-digit balance. The density of each specimen was calculated by using the following equation:

$$D = \frac{M}{V} \tag{1}$$

where D is density of specimen (g/cm^3), m is weight (g) and V is volume (cm^3).
The weight per unit area of paper sheets was defined as grammage (according to TAPPI T410). The grammage was calculated from the ratio of the mass (g) to the area of sheets (m^2). The porosity of each specimen was calculated by using the following equation:

$$\%Porosity = \left(1 - \frac{density\, of\ sheet}{density\, of\ cellulose}\right) \times 100 \tag{2}$$

where the density of cellulose is 1.53 g/cm^3.

IOP Conf. Series: Materials Science and Engineering **559** (2019) 012007 doi:10.1088/1757-899X/559/1/012007

2.4.3. Tensile test. To prepare the test specimens, the paper sheets were cut into the size of 10 mm × 50 mm with a gauge length of 30 mm. Then, the specimens were pre-conditioned at 50% RH and 25°C overnight before testing. The tensile test was performed according to TAPPI T494 by using a universal testing machine (Instron Model 5566) equipped with 1 kN load cell. The constant crosshead speed of 15 mm/min and pre-load at 1 N were set before starting the test. The tensile index, tensile strength, breaking length and Young's modulus of the prepared paper sheets were evaluated subsequently from the test results. The reported values of their tensile properties were averaged from 5 specimens.

3. Results and Discussion

3.1. Morphology

Figure 1A shows the SEM micrograph of BG fibers with their dimensions in microscale level (see also Table 1). On the other hand, after passing through the microfluidizer for 10 cycles, BC network was turned into individual nanofibers. However, some were still together and remained partially network-like (Figure 1B) [5]. From Table 1, it can be clearly noticed that the BG fibers have a considerably low aspect ratio when compared to that of the BC nanofibers.

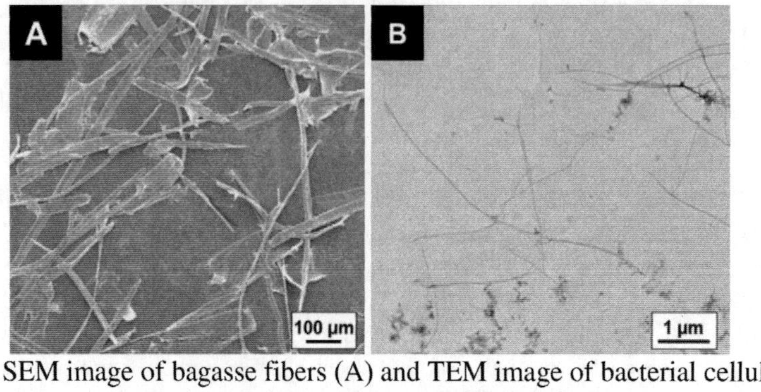

Figure 1. SEM image of bagasse fibers (A) and TEM image of bacterial cellulose (B) after 10 passes through the microfluidizer.

Table 1. Size of bagasse microfibers (BG) and bacterial cellulose (BC) nanofibers

Type of fiber	Length (L)	Width (D)	Aspect ratio (L/D)
BG	274.8 ± 105.9 μm	11.7 ± 4.9 μm	29
BC	10.5 ± 2.7 μm	3.5 ± 2.0 nm	336

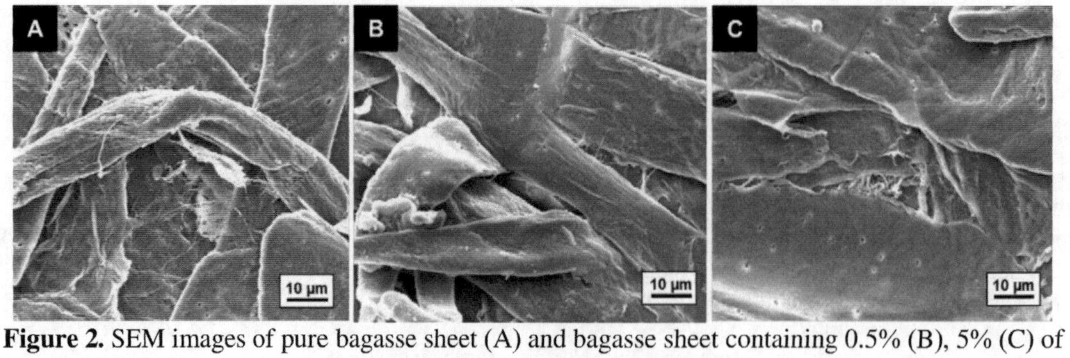

Figure 2. SEM images of pure bagasse sheet (A) and bagasse sheet containing 0.5% (B), 5% (C) of bacterial cellulose at 1000x magnification.

The SEM images of the surfaces of the pure BG sheet and BG sheets contained BC are shown in figure 2A and 2B-C, respectively. It was revealed that the added BC nanofibers filled the gaps between BG microfibers and smoothen the sheet surfaces (figure 2B-C).

3.2. Properties of the paper sheets

The plot of the density and porosity of the prepared paper sheets is shown in figure 3. It was found that the density of the paper sheets was increased with increasing BC content, but their porosity was steadily decreased. It was likely due to BC filling the gaps in the BG sheet structures and, hence, the less porous sheets with higher density were obtained as a result [6]. In general, the density and porosity of papers are highly related to their mechanical properties. It is supposed that the higher density papers have greater fibers packing, which could lead to an enhancement in the mechanical properties of paper sheets [7].

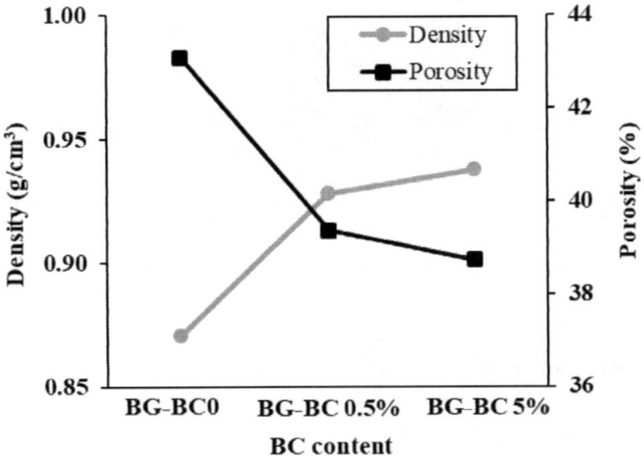

Figure 3. Density and porosity of the pure BG, and BG with 0.5% and 5% BC paper sheets.

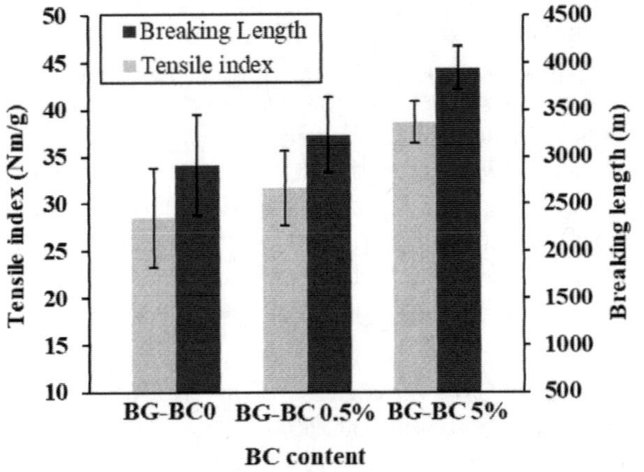

Figure 4. Reinforcing effect of BC nanofibers on tensile index and breaking length of the BG sheets.

From the tensile results, the addition of BC improved tensile properties of the BG paper sheets as shown in Figure 4 and 5. Because the presence of BC increased the bonding ability within the sheet structures and they closed the gaps by bridging between these BG fibers [3,6]. For the BG sheet with only 0.5% BC, its tensile index and breaking length were increased approximately 10%. When increasing BC content to 5%, the tensile index and breaking length of the sheet were further enhanced up to 35% increment (figure 4). In the previous work, T. Tabarsa et al. studied the use of BC to improve properties of the softwood pulp (SP). The 5% BC reinforced SP sheet showed an increase in tensile index of 12% [6]. C. Campano et al. also reported that the recycled pulp sheet with 6% BC addition presented an increase in its tensile index around 22% [8]. From the comparison with these former works, it was indicated that the present defibrillated BC fibers showed a greater reinforcing efficiency than unmodified or untreated BC.

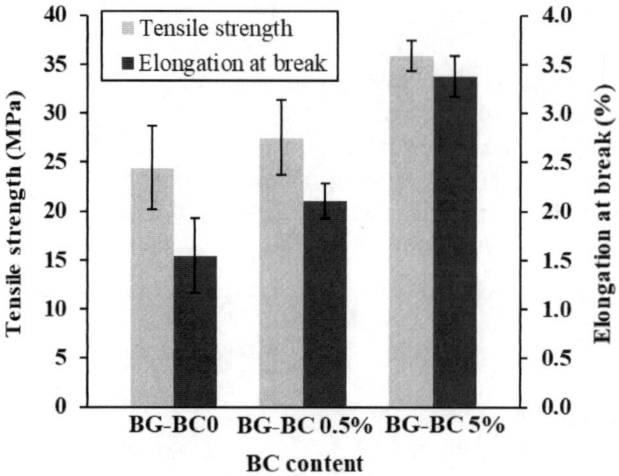

Figure 5. Reinforcing effect of BC nanofibers on tensile strength and elongation of the BG sheets.

In addition, the 5% BC added sheet (BG-BC 5%) showed the significant increase in its tensile strength and elongation at break (47% and 117%, respectively) when compared to those of the pure sheet (BG-BC0) (figure 5). The overall results also confirmed that the defibrillated BC nanofibers were able to effectively improve all tensile properties including tensile index, breaking length, tensile strength, and elongation at break of the BG paper sheets.

4. Conclusion

The defibrillated BC nanofibers were prepared by using a microfluidizer. Different contents (0.5 and 5 wt%) of BC were incorporated in the BG pulp and then fabricated into the paper sheets by using a compression molding. As the dosage of BC in the BG sheets was increased, their density was raised but porosity was reduced because BC was found to fill in the gaps and bridge between BG fibers as revealed by SEM images. With 5 wt% BC addition, the tensile index and breaking length of the BG sheet were improved by approximately 35%. Its tensile strength and elongation at break were also greatly enhanced by 47% and 117%, respectively. The potential reason for these improvements attributes to an increase in inter-fiber bonding within the BG sheet structure due to the presence of BC nanofibers. Thus, it can conclude that the defibrillated BC is an effective choice and feasible to use for developing papers, specialty papers and other related products.

5. References

[1] Shah N, Ul-Islam M, Khattak 2013 *Carbohydr Polym.* **98** 1585
[2] Basta A H, El-Saied. *J Appl Microbiol.* **107** 2098
[3] Yuan J, Wang T, Huang X 2016 *BioResources* **11** 9290
[4] Khakifirooz A, Ravanbakhsh F 2012 *BioResources*. **8** 21
[5] Afra E, Yousefi H, Hadilam M M 2013 *Carbohydr Polym.* **97** 725
[6] Tabarsa T, Sheykhnazari S 2017 *Int J BiolMacromol.* **101** 334
[7] Qing Y, Sabo R, Zhu J Y 2013 *Carbohydr Polym.* **97** 226
[8] Campano C, Merayo N 2018 *Int J BiolMacromol.* **114** 1077

Acknowledgments

The joint funding support (grant code 256101A3070019) from the National Research Council of Thailand (NRCT) and National Science Technology and Innovation Policy Office (STI) is gratefully acknowledged. All supports and facilities from our industrial partner, the Biodegradable Packaging for Environment Public Co., Ltd. (Gracz) as well as Mae Fah Luang University (MFU) and the PLC Holding Co., Ltd., Thailand are also much appreciated.

IOP Conf. Series: Materials Science and Engineering **559** (2019) 012008 doi:10.1088/1757-899X/559/1/012008

Effect of Pineapple Leaf Nanofibers on Physical and Mechanical Properties of Bagasse Sheets

N Kunfong[1], P Rattanawongkun[1], N Tawichai[1,2], S Duangphet[1,2] U Intatha[1,2], and N Soykeabkaew [1,2]

[1]School of Science, Mae Fah Luang University,333 M 1, Muang, Chiang Rai 57100, Thailand
[2]Center of Innovative Materials for Sustainability (iMatS), Mae Fah Luang University, 333 M 1, Muang, Chiang Rai 57100, Thailand

Email: nattakan@mfu.ac.th

Abstract. Agro-waste pineapple leaves (PA) were extracted into cellulose nanofibers by pretreated with soda pulping process and then disintegrated by microfluidizer. The resulting PA nanofibers with an average length of 1.7±0.5 μm and diameter of 15.5±3.6 nm was then used to improve properties of bagasse (BG) pulp sheets at different contents (0.5, 1, and 5 wt%). The results showed that integration of PA nanofibers increased the density of the sheets but decreased their porosity. With addition of 5 wt% PA nanofibers, BG sheets showed the increases in all tensile properties, especially tensile strength (64%) and elongation (140%). SEM images revealed the smoother surface of the sheets and indicated the reduction of the pores by PA nanofibers filling and bridging between BG fibers network. This increased fibers bonding area in the sheet structures and hence improving their mechanical properties.

1. Introduction
There has been an intensive study of paper reinforcement by adding additives and cellulose nanofibers. Since cellulose nanofibers have high aspect ratio, surface area, and crystallinity etc. [1], it can improve fibers bonding and reduce pores between fibers leading to an enhancement in the air resistance and mechanical properties of papers [2]. Cellulose nanofibers can be extracted from both wood and non-wood resources such as eucalyptus, coconut husk, oil palm, rice straw, bamboo, bagasse, and pineapple leaf [3]. Agro-waste pineapple leaf is abundant in South East Asian region [4], relatively inexpensive and has high specific strength, stiffness and crystallinity (70 – 82%) [5]. Therefore, it is considered as an outstanding alternative raw material to produce cellulose nanofibers. However, no research in using the pineapple leaf nanofibers as reinforcement in paper has been reported so far. This research aims to study the effect of adding PA nanofibers on physical and mechanical properties of paper made from bagasse pulp.

2. Experimental

2.1. Materials
Bleached bagasse pulp in dry sheet form supplied by Biodegradable Packaging for Environmental Co., Ltd (BPE), Thailand was used as the raw material. Nanglae pineapple leaves used in this study were obtained from the cultivation area in Chiang Rai, Thailand. Sodium hydroxide (NaOH) AR grade pellet was supplied by QRëC™.

Content from this work may be used under the terms of the Creative Commons Attribution 3.0 licence. Any further distribution of this work must maintain attribution to the author(s) and the title of the work, journal citation and DOI.
Published under licence by IOP Publishing Ltd

IOP Conf. Series: Materials Science and Engineering **559** (2019) 012008 doi:10.1088/1757-899X/559/1/012008

2.2. Preparation of cellulose nanofibers from pineapple leaf
Pineapple leaves (PA) were cut into 10 cm and dried until constant weight. The dried PA were then ground into small pieces. After that, they were cooked in 18%(w/w) NaOH at a controlled temperature of 98 ± 2°C for 30 minutes. The PA pulp was washed to remove the unreacted chemical reagent and then blended by the shear mixer at speed 15,000 rpm for 1 minute. Mesh no.18 (1 mm) and no.200 (0.074 mm) were used to screen the pulp. To produce the nanofibrils, the pulp suspension was diluted to 0.5% consistency and refined in the microfluidizer (M-110P, Microfluidics). The pulp was passed through the microfluidizer with an 87 μm chamber under pressure of 25,000 psi for 20 cycles.

2.3. Paper sheet preparation
Bleached bagasse (BG) sheet form was torn into small pieces and soaked in water overnight. Then, it was disintegrated by shear mixing 15,000 rpm in a household blender for 10 minutes. The BG pulp that passed through mesh no.18 but contained in mesh no.200 was collected. The screened BG pulp was diluted to 2% consistency. The certain amount (0.5, 1 and 5 wt%) of nanofibers was added to BG pulp. The paper sheets were prepared by using hot press compression molding technique. They were pressed under pressure of 159 kPa and heated at 105°C for 5 minutes.

2.4. Morphological studies
The surface morphology of BG and different amount of nanofibrils reinforced sheets was observed by scanning electron microscope (SEM) LEO/1450 VP operating at an accelerating voltage of 10 kV. The carbon was coated onto samples surface before observation. The PA nanofibers were investigated by transmission electron microscope (TEM, Hitachi model HT7700). The fibers were diluted to 0.001% concentration and drop on a copper grid with the formvar film. The sample was dyed with 2% uranyl acetate for 5 minutes and performed at an accelerating voltage of 80 kV.

2.5. Physical properties
The grammage of the sheet expressed in terms of grams per square meter (g/m^2) according to TAPPI T410. The density of the sheets was calculated from the ratio of the mass to the area by measuring the dimension and weight.

2.6. Mechanical properties
Before testing, each sheet was conditioned at 25°C and 50% RH at least overnight. Mechanical properties of the sheets including tensile index, Young's modulus, elongation and breaking length were calculated from the results of tensile test. Tensile testing was performed by the universal testing machine (Instron Machine model 5566) with 1 kN cell load (according to TAPPI T494). The gauge length of the sample was set at 30 mm with the width 10 mm.

3. Results and Discussion
The SEM and TEM were used to examine the morphology of the BG fibers and PA nanofibers. The SEM micrograph of BG shows in Figure 1a. the dimension of BG fiber was reported in average value from 50 fibers. The length and diameter of BG fibers is 275 ± 11 μm and 12 ± 5 μm, respectively. The TEM image of PA nanofibers shows in Figure 1b. The dimension of PA nanofibers was reported in average value from 10 fibers. The length and diameter of PA nanofibers is 1.7 ± 0.5 μm and 15.5 ± 3.6 nm, respectively.

The physical properties including grammage, density, and porosity of the pure BG sheet and BG sheet reinforced by the PA nanofibers were summarized in Table 1. The grammage of both the BG sheets and BG sheets reinforced by PA nanofibers were not significantly different. The density of the sheets increases as the amount of PA nanofibers increases but in contrast to the porosity. It is because PA nanofibers can fill up the gap between fibers [2] and nano-network was created between fibers [6]. Nano-network created higher bonding and packing between fibers [7] as shown in figure 2.

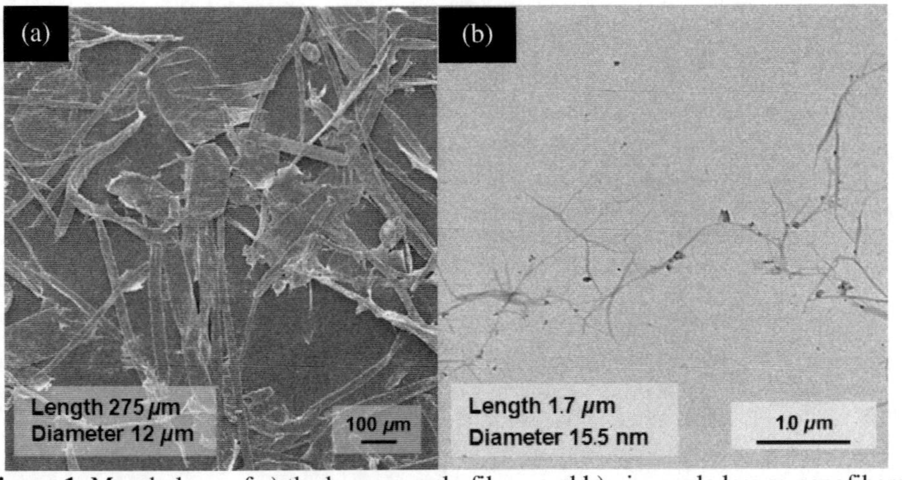

Figure 1. Morphology of a) the bagasse pulp fibers and b) pineapple leaves nanofibers.

Table 1. Physical properties of the BG sheets at different content of adding PA nanofibers.

Sample	Grammage (g/m^2)	Density (g/cm^3)	Porosity (%)
BG	288.30	0.87	43.10
BG-PA 0.5%	254.95	0.94	38.64
BG-PA 1%	294.17	1.03	32.80
BG-PA 5%	275.45	1.06	30.96

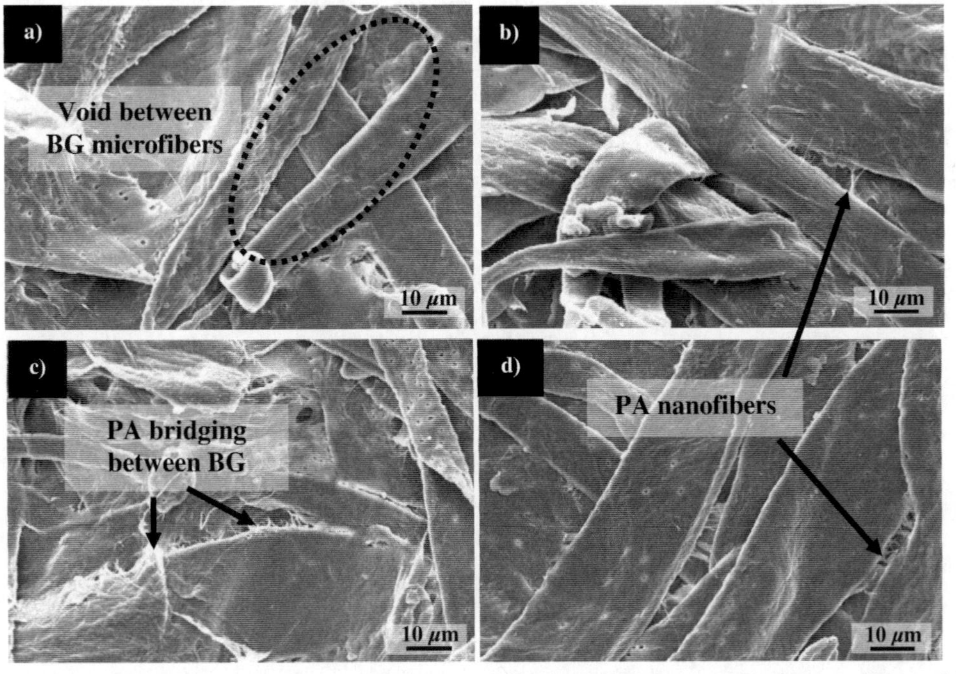

Figure 2. SEM images taken under 1,000× magnification of bagasse (BG) sheets reinforced by different contents of pineapple leaves nanofibrils (PA). a) 0wt%; b) 0.5wt%; c) 1wt%; d) 5wt%.

Figure 2. shows the surface of the BG sheet and BG reinforced by PA nanofibers sheets. Figure 2a. displays the voids between BG fibers and no network between fibers when compared to the sheets added with 0.5, 1 and 5 wt% content of PA nanofibers. The addition of PA created the connection or bridging of this nanofiber and reduced the gaps between BG fibers (Figure 2b-d). Moreover, it was

IOP Conf. Series: Materials Science and Engineering **559** (2019) 012008 doi:10.1088/1757-899X/559/1/012008

observed that the sheets tended to be smoother when added more nanofibers. The 5wt% PA nanofiber reinforced sheet showed the smoothest surface. Furthermore, it was found that the mechanical properties of the sheets were also considerably enhanced by adding the PA cellulose nanofibrils as presented in Table 2.

Table 2. Mechanical properties of the BG sheets at different contents of adding PA nanofibers.

Sample	Tensile strength (MPa)	Elongation (%)	Young's modulus (MPa)	Breaking length (m)
BG	24.42 ± 4.26	1.55 ± 0.39	3060.20 ± 592.04	2914.01 ± 535.10
BG-PA 0.5%	28.48 ± 2.41	2.53 ± 0.42	2838.20 ± 130.37	3226.30 ± 602.01
BG-PA 1%	30.66 ± 3.64	2.00 ± 0.29	3348.90 ± 288.59	3361.36 ± 803.04
BG-PA 5%	40.05 ± 4.63	3.72 ± 0.49	3450.58 ± 262.34	3794.04 ± 563.75

The average tensile strength, elongation, Young's modulus, and breaking length of all sheets were summarized in Table 2. The results indicated that the mechanical properties highly depend on the amount of cellulose nanofibrils content. The BG sheet with adding PA nanofibers 5wt% shows the maximum all of the mechanical properties followed by 1, 0.5, and 0 wt% of PA nanofibers content. The results shown the tensile strength, elongation, Young's modulus and breaking length increase to 64%, 140%, 13%, and 30% respectively when compared to the sheet with 0wt% PA nanofibers addition. The highest elongation from adding PA nanofibers 5wt% to BG sheet indicates to high inter-fiber bonding area in the sheets [8]. Elongation is measured by the strain at breaking point as can be seen in stress-strain curves of the sheets from the tensile testing (see Figure 3).

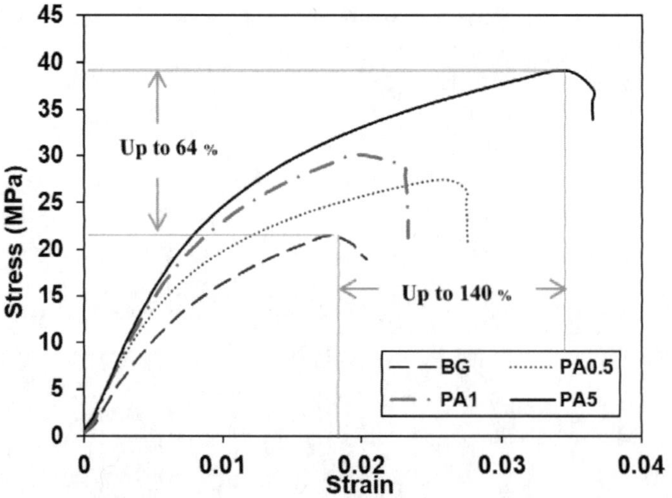

Figure 3. Stress-strain curves of BG sheets with different amounts of cellulose nanofibers.

The superior mechanical properties of this current sheets were contributed from reinforcing effect of cellulose nanofibers which fulfill the pore between fibers and better internal bonding of the sheet [1, 9, 10]. Similarly, Ding et al. [11] investigated that the fluorescent cellulose nanofibrils (FCNF) had high fiber-bond degree with the pulp fibers and highly enhanced the mechanical properties of papers. But its elongation increased only 81% with adding 10wt% of FCNF to the paper.

4. Conclusion

The effect of different PA nanofibers content on physical and mechanical properties of bagasse sheet was investigated. The higher content of PA nanofibers to BG sheets resulted in increased density, reduced porosity and enhanced tensile properties of the BG sheets. Addition of PA nanofibers at 5wt%

on BG sheet showed the increases in tensile strength, elongation, Young's modulus and breaking length to 64%, 140%, 13% and 30% respectively. The notable positive effect on mechanical properties of BG sheets involved the reduction of the gap between fibers due to the filling of PA nanofibers which also increased bonding area within the sheets structure.

5. References

[1] Mtibe A, Linganiso L Z 2015 *Carbohydr Polym.* **118** 1
[2] Tajik M, Torshizi H J, Resalati 2018 *Carbohydr Polym.* **194** 1
[3] Laftah W A, Rahaman W A 2015 *Mater Res Technol.* **4(3)** 254
[4] Mahardika M, Abral H, Kasim A, Arief S 2018 *M. MDPI.* **6** 28
[5] Asim M, Abdan K 2015 *Int J Polym Sci.* **2015** Article ID 950567
[6] Balea A, Blanco Á, Monte M C 2016 *BioResources.* **11(4)** 8123
[7] Yuan J, Wang T, Huang X 2016 *BioResources.* **11(4)** 9290
[8] Tabarsa T, Sheykhnazari S 2017 *Int J Biol Macromol.* **101** 334
[9] Rantanen J, Dimic-Misic K 2015 *Cellulose.* **22** 4003
[10] Afra E, Yousefi H, Hadilam 2013 *Carbohydr Polym.* **97** 725
[11] Ding Q, Zeng J, Wang B, Gao W 2018 *Carbohydr Polym.* **186** 73

Acknowledgments

The joint funding support (grant code 256101A3070019) from the National Research Council of Thailand (NRCT) and National Science Technology and Innovation Policy Office (STI) is gratefully acknowledged. All supports and facilities from our industrial partner, the Biodegradable Packaging for Environment Public Co., Ltd. (Gracz) as well as Mae Fah Luang University (MFU) and the PLC Holding Co., Ltd., Thailand are also much appreciated.

IOP Conf. Series: Materials Science and Engineering **559** (2019) 012009 doi:10.1088/1757-899X/559/1/012009

Electronically Adjustable Voltage-mode First-order Allpass Filter Using Single Commercially Available IC

Amornchai Chaichana, Surapong Siripongdee and Winai Jaikla

Department of Engineering Education, Faculty of Industrial Education and Technology, King Mongkut's Institute of Technology Ladkrabang, Bangkok, Thailand, 10520

amornchai.ch@kmitl.ac.th

Abstract. In this design, the use of single commercially available integrated circuit (IC), LT1228 from Linear Technology Corporation, to implement the first order allpass filter is presented. The passband voltage gain of the presented first allpass filter (APF) is unit and constant throughout the operation frequency, while the phase response is electronically tuned by changing the external DC bias current. With this property, the proposed first order filter can be used as phase shifter with microcontroller controllability. Tuning phase property is not required any matching condition. The proposed phase shifter consists of one LT1228, one capacitor and two resistors which is attractive for off the shelf implementation. Moreover, the impedance at output voltage node is exhibit low which can connect to other circuit without the requirement of external voltage buffer. The expected performances of the proposed circuit are proved by Pspice simulation using macro model parameters. Also, the quadrature sinusoidal oscillator which is designed from the connecting of the proposed filter and integrator is chosen as application example.

1. Introduction

It is well-known that analog filter is one of the important functions in analog signal processing circuits. This circuit is found in many applications for examples in communication system, power electronic systems, instrument and measurement system etc. [1]. The allpass filter is one of analog filter which has been developed and proposed continuously [2-5]. The property of allpass filter is different from the high-pass, low-pass, band-pass and band-stop filter because there is no stop-band for allpass filter. So, the allpass filter is used to pass the input signal to output signal with constant amplitude for all frequency. However, the phase of the output signal will be shifted from the phase of input signal [6].

The LT1228 is integrated circuit manufactured by Linear Technology Corporation [7]. This integrated circuit is considered as active building block (ABB) which is interesting active device for design the analog signal circuit. The LT1228 consists of two basic active blocks, operational transconductance amplifier (OTA) and current feedback amplifier (CFA). The transonductance (g_m) of LT1228 is electronically controlled via DC bias current. So, the LT1228 based circuits [8-11] are easily controlled by microcontroller which is the important requirement for modern circuit.

Several voltage-mode first order allpass filters or phase shifter circuits employing the voltage or current ABBs have been reported in the open literature [12-31]. However, these reposed first order allpass filters have at least one of the following inconveniencies:

- Contain more than one active building block [13, 18, 21].

Content from this work may be used under the terms of the Creative Commons Attribution 3.0 licence. Any further distribution of this work must maintain attribution to the author(s) and the title of the work, journal citation and DOI.
Published under licence by IOP Publishing Ltd

- Lack of electronic controllability [12, 17, 23, 24, 26, 27, 28, 29, 30, 31].
- Tuning phase property requires the constraint matching condition of two active or passive elements [15, 16, 17, 25, 28, 29].
- Impedance at output voltage node doesn't exhibit low which requires the voltage buffer for cascading [15, 16, 17, 18, 19, 20, 21, 22, 23, 24, 25, 27, 28, 29, 30, 31].
- Need the active building block which contains multiple current output terminal [15, 16, 24, 27].
- The used active building block is not implemented from single commercially available IC [14, 15, 16, 17, 18, 19, 20, 23, 24, 26, 27, 28, 29, 30].

This paper introduces the design of voltage mode phase shifter circuit employing single LT1228 as active building block connecting with single capacitor and two resistors. The tuning of phase response is electronically adjusted with no need of the constraint matching condition of two active or passive elements. The functionalities of the presented phase shifter circuit were simulated by Pspice program.

2. Principle of Operation

In this section, the details of active building block and proposed first order allpass filter are given as follows.

2.1. LT1228

In this section, brief detail of the main active device will be described. The internal configuration of LT1228 is shown in Fig. 1. It has eight pins which is the connection of OTA and CFA. The high impedance inverting and non-inverting voltage inputs are pin 2 and 3, respectively. The high impedance current output of OTA which is connected to the non-inverting voltage input of CFA is at pin 1. The external bias current used to control the transconductance of OTA is at pin 5. The high impedance non-inverting voltage input of CFA is at pin 8. The low impedance voltage output is at pin 6. The negative and positive voltage power supplies are respectively at pin 4 and pin 7. The electrical symbol of LT1228 is illustrated in Fig. 2. The electrical property of LT1228 is characterized with the following matrix equation

$$I_Z = g_m\left(V_+ - V_-\right);\ V_x = V_z;\ V_W = I_X R_T \tag{1}$$

where R_T represents the transresistance of LT1228 and it is infinity in ideal case. The transconductance gain of LT1228 is adjusted electronically via I_B [3]

$$g_m = 10 I_B \tag{2}$$

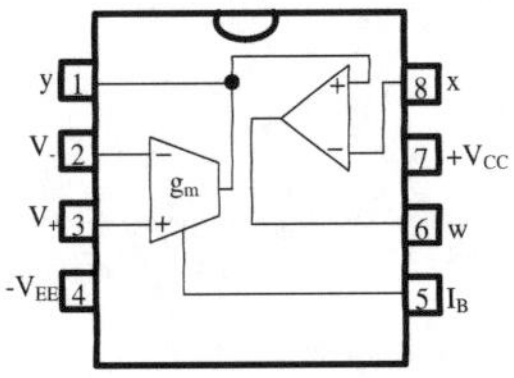

Figure 1. Pin configuration of LT1228. [7]

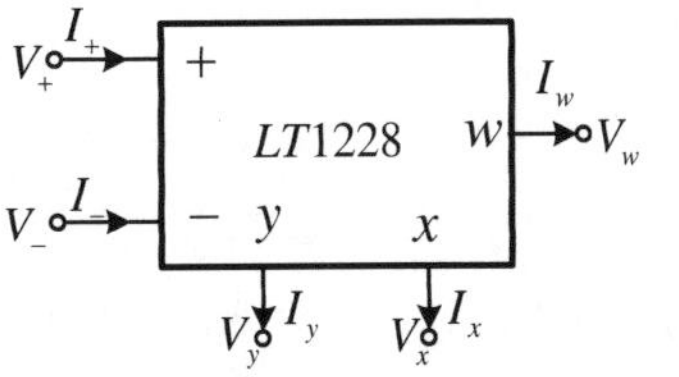

Figure 2. Symbolic of LT1228. [11].

2.2. Proposed voltage-mode first order allpass filter

Figure 3 shows the proposed circuit composed of one LT1228, one capacitor and two resistors. It is found that the voltage output is at low impedance terminal w. The voltage transfer function of the circuit in Fig. 3 is as follow:

$$T(s) = \frac{v_o}{v_i} = -\left(\frac{g_m - sC}{g_m + sC}\right) \tag{3}$$

Considering the voltage transfer function as appeared in (3), passband voltage gain, the phase response (θ) and the natural frequency (ω_0) are obtained respectively as

$$|T(s)| = 1; \theta = 180 - 2\tan^{-1}\left(\frac{\omega C}{g_m}\right) \text{ and } \omega_0 = \frac{g_m}{C} \tag{4}$$

Thus, the θ and ω_0 are electronically varied by adjusting the external DC bias current I_B. It should be noted from (4) that the phase response is shifted from 180° to 0° which is implied that the phase of the output voltage is leaded to the phase of the input voltage. Moreover, the ω_0 can be linearly controlled via I_B.

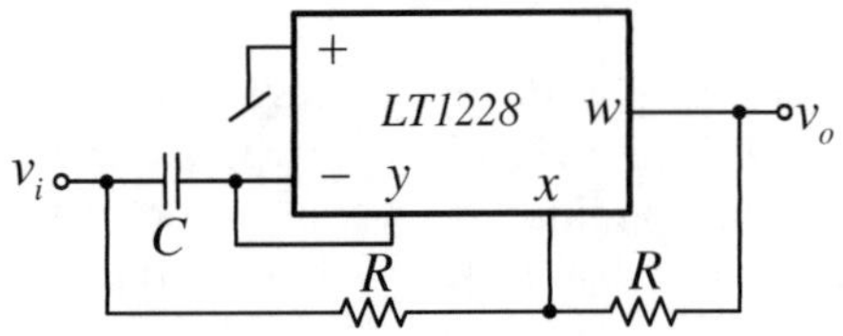

Figure 3. The proposed first order allpass filter

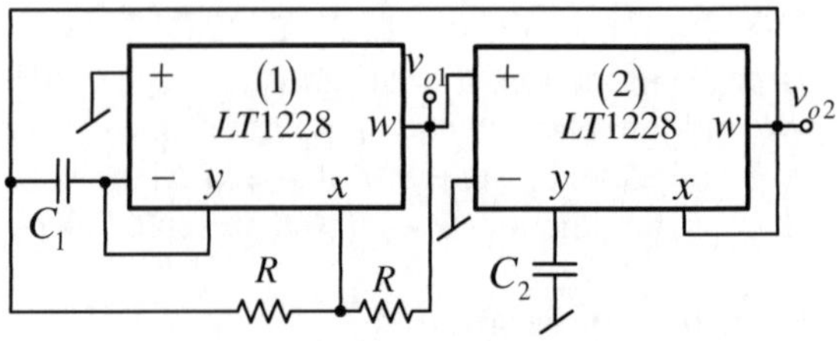

Figure 4. The quadrature oscillator based on proposed first order allpass filter.

2.3. Application example as the quadrature sinusoidal oscillator

To show the utilization of the proposed fist order filter, it is employed to design the quadrature oscillator by connecting the first order and inverting lossy integrator. The complete voltage mode quadrature sinusoidal oscillator is shown in Fig. 4. It is noted that the output voltage nodes, v_{o1} and v_{o2} are at w terminals which have low impedance. Consequently, it can be connected to other circuit or driven external load without the requirement of any external voltage buffer. Routine analysis of the circuit in Figure 4, the characteristic equation is obtained as

$$s^2C_1C_2 + sC_2g_{m1} - sC_1g_{m2} + g_{m1}g_{m2} = 0 \tag{5}$$

From (5), the frequency of oscillation (FO) and condition of oscillation (CO) can be expressed as

$$\omega_0 = \sqrt{\frac{g_{m1}g_{m2}}{C_1C_2}} \text{ and } C_2g_{m1} \le C_1g_{m2} \tag{6}$$

It is found from (6) that the FO and CO can be electronically controlled via g_{m1} or g_{m2} which is easily controlled by microcontroller. The phase relationship of v_{o1} and v_{o2} is given by

$$\frac{v_{o2}}{v_{o1}} = \frac{g_{m2}}{sC_2} \tag{7}$$

The phase difference between the output voltage v_{o1} and v_{o2} is 90° where the phase of the v_{o1} signal is leaded to the phase of the v_{o2} signal as analyzed in (7).

3. Simulation Results

First, the workability of the filter topology given in Fig. 3 was verified through PSPICE program. The DC power supply voltages, $V_{CC} = -V_{EE}$ were equal to ±5V and external DC bias current of LT1228 employed to adjust the transconductance value was chosen as I_B = 65 μA. In this design, the values of resistors and capacitor were chosen as 1kΩ and 1nF, respectively. The simulated passband voltage gain and phase characteristic of the output voltage are shown in Fig. 5. It is evident that the passband voltage gain is unity (0dB) and constant over operational frequency while phase response is shifted from 180° to 0° which is accordance to the theoretical prediction (Eq. 4). At frequency of 100kHz, the simulated values of passband voltage gain and phase response are 0.992 and 91.52° respectively. The theoretical these values calculated from (4) at frequency of 100kHz are 1 and 91.97°, respectively. A little error deviation of 0.8% and 0.49% between the theoretical and simulation value stems from the non-ideality of LT1228 such as the voltage/current tracking errors and the parasitic elements etc.

Figure 6 shows the phase response of the output voltage where the external DC bias current I_B was changed to 35μA, 65μA and 115μA. It is evident that at frequency of 100kHz, the phase response are respectively located to 57.49°, 91.52° and 121.99°, respectively. This result confirms the theoretical expectation as analyzed in (4).

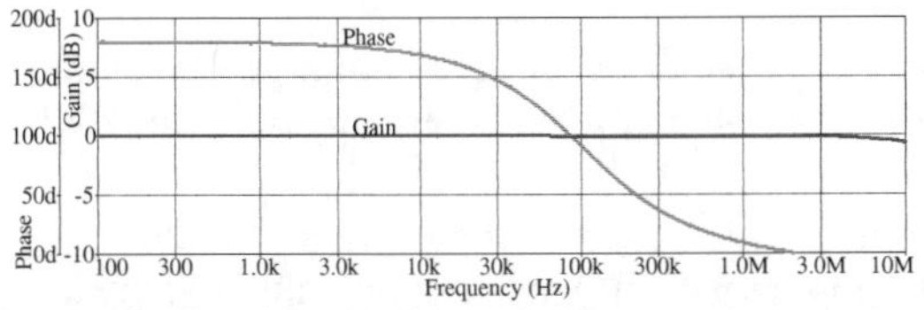

Figure 5. Simulated gain and phase response

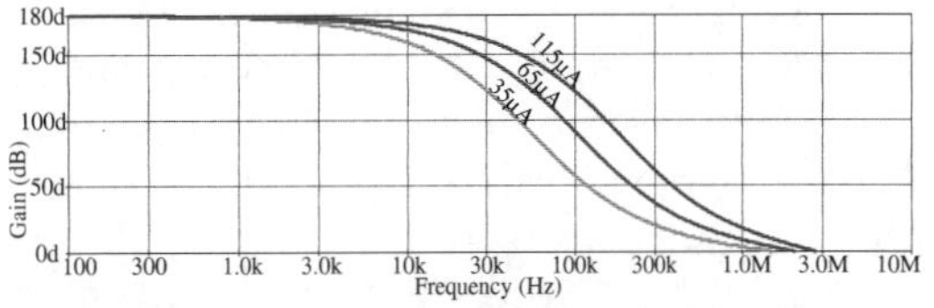

Figure 6. Phase relationship of for different I_B.

Second, the performances of the quadrature sinusoidal oscillator in Figure 4 were verified via PSpice program. The DC power supply voltages, $V_{CC} = -V_{EE}$ were equal to ±5V and external DC bias currents of LT1228 employed to adjust the transconductance values were chosen as I_{B1} = 65 μA and I_{B2}=70μA. In this design, the values of resistors and capacitor were chosen as R=1k and $C_1=C_2$=1nF. The simulated sinusoidal output waveform during steady state is shown in Fig. 7. It is found that the circuit in Fig. 4 can generated the quadrature sinusoidal output waveform as analyzed in (7). The simulated output spectrum of v_{o1} and v_{o2} signals is illustrated in Fig. 8, where the FO is 100kHz. The total harmonic distortions (THD) of v_{o1} and v_{o2} signal are 1.54% and 1.31%, respectively.

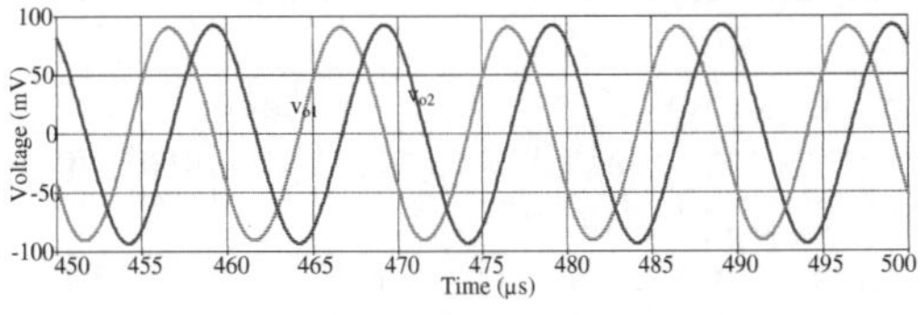

Figure 7. Simulated quadrature output waveform

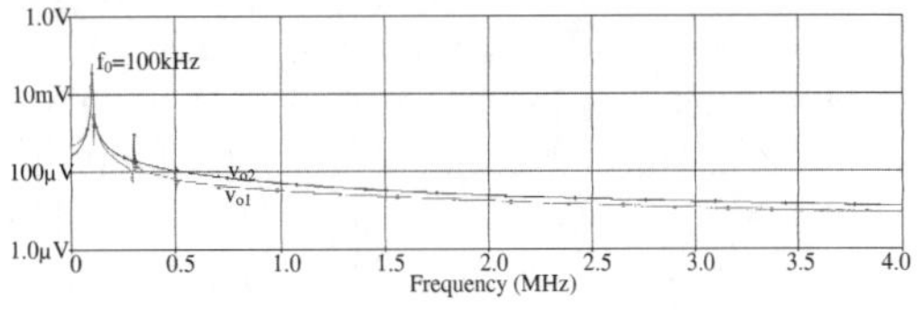

Figure 8. Output spectrum.

4. Conclusion

The single commercially available IC, LT1228 based voltage-mode first order allpass filter is designed in this paper. The advance features of the proposed circuit are electronic tunable phase response, low output impedance and minimum number of active element. Moreover, the matching condition of active or passive element is not required for tuning the phase response which is easy to design the bias current circuit by controlling with microcontroller. To show the usability of the proposed circuit, the quadrature oscillator is designed by connecting the proposed first order filter and lossless integrator. The presented circuits were verified via Pspice simulations and the results agree well with theoretical expectation.

5. References

[1] C. Psychalinos, C. Kasimis, and F. Khateb, "Multiple-input single-output universal biquad filter using single output operational transconductance amplifiers," AEU - International Journal of Electronics and Communications, vol. 93, pp. 360-367, 2018.

[2] W. Jaikla, A. Noppakarn and S. Lawanwisut, "New gain controllable resistor-less current-mode first order allpass filter and its application," Radioengineering, vol. 21, no. 1, pp. 312–316, 2012.

[3] A. Chaichana, M. Kumngern, M. Siripruchyanun, and W. Jaikla, "Amplitude controllable current-mode first order allpass filter including minimum component count circuits," Proceedings of 38th International Conference on Telecommunications and Signal Processing, 2015.

[4] W. Tangsrirat, "On the realization of first-order current-mode AP/HP filter," Radioengineering, vol. 22, pp. 1007-1015, 2013.

[5] W. Tangsrirat, T. Pukkalanun, and W. Surakampontorn, "Resistorless realization of current-mode first-order allpass filter using current differencing transconductance amplifiers," Microelectronics Journal, vol. 41, pp. 178-183, 2010.

[6] D. Biolek and V. Biolkova, "Allpass filter employing one grounded capacitor and one active element," Electronics Letters, vol. 45, pp. 807-809, 2009.

[7] http://www.linear.com/product/LT1228

[8] S. Klungtong, D. Thanapatay, and W. Jaikla, "Three-input single-output voltage-mode multifunction filter with electronic controllability based on single commercially Available IC," Active and Passive Electronic Components, 2017.

[9] S. Siripongdee and W. Jaikla, "Universal filter using single commercially available IC: LT1228," Proceedings of the 3rd International Conference on Mechatronics and Mechanical Engineering, 2016.

[10] S. Siripongdee and W. Jaikla, "Electronically controllable grounded inductance simulators using single commercially available IC: LT1228," AEU – International Journal of Electronics and Communications, vol. 76, pp. 1-10, 2017.

[11] P. Suwanjan, S. Siripongdee, and W. Jaikla, "Three-inputs single-output voltage-mode universal filter with orthogonal control using single commercially available IC," Proceedings of the 2017 European Conference on Electrical Engineering and Computer Science (EECS), 2017, pp. 454-457.

[12] R. Nandi, K. Mathur, and S. Pattanayak, "Single-CFA first-order allpass filter," IEICE Electronics Express, vol. 13, 2016.

[13] R. Nandi, K. Mathur, and S. Pattanayak, "Electronically tunable allpass filter: linear VCO design," IEICE Electronics Express, vol. 13, 2016.

[14] J. Satansup, and W. Tangsrirat, "Single VDGA-base first-order allpass filter with electronically controllable passband gain," Proceedings of the 7th International Conference on Information Technology and Electrical Engineering, ICITEE 2015, 2015, pp. 106–109.

[15] D. Prasad, K. Panwar, D.R. Bhaskar, and M. Srivastava, "CDDITA-based voltage-mode first order all pass filter configuration," Circuits and systems, vol. 6, pp. 252–256, 2015.

[16] S. Maneewan, N. Udorn, D. Duangmalai, P. Silapan, and W. Jaikla, "A voltage-mode first order allpass filter based on VDTA," Advances in Electrical and Electronic Engineering, vol. 12, pp. 40–46, 2014.

[17] J. W. Horng, C. M. Wu, and N. Herencsar, "Fully differential first-order allpass filters using a DDCC," Indian Journal of Engineering and Materials Sciences, vol. 21, pp. 345-350, 2014.

[18] V. K. Dixit, R. Gupta, and K. Pal, "Novel DVCCs based voltage-mode first-order allpass section," Proceedings of the 2014 International Conference on Signal Processing and Integrated Networks, SPIN 2014, pp. 497–501, 2014.

[19] M. Kumngern, V. Mongkol and S. Junnapiya, "Voltage-mode allpass section employing only one DDCCTA and one capacitor," Proceedings of the Eleventh International Conference on ICT and Knowledge Engineering, pp. 1–4, 2013.

[20] S.Z. Iqbal, C. Psychalinos and N. Parveen, "First-order allpass filter using multi-input OTA," International Journal of Electronics, vol. 100, no. 10, pp. 1373-1382, 2013.

[21] M. Kumngern, "Realization of electronically tunable first-order allpass filter using single-ended OTAs," Proceedings of the IEEE Symposium on Industrial Electronics and Applications, ISIEA 2012, vol. 12, pp. 100–103, 2012.

[22] M. Kumngern, and J. Chanwutitum, "Voltage-mode first-order allpass filter using single-ended OTAs and grounded capacitor," Proceedings of the Second International Conference on Digital Information and Communication Technology and its Applications, DICTAP 2012, pp. 305–308, 2012.

[23] H. Chen, S. Wang and K. Huang , "Grounded-capacitor first-order filter employing single DVCC," Proceedings of the IEEE 3th International Conference on Communication Software and Networks, vol. 11, pp. 401–404, 2011.

[24] J. W. Horng, "High input impedance first-order allpass, highpass and lowpass filters with grounded capacitor using single DVCC," Indian Journal of Engineering and Materials Sciences, vol. 17, pp. 175-178, 2010.
[25] C. Tanaphatsiri, W. Jaikla, and M. Siripruchyanun, "An electronically controllable voltage-mode first-order all-pass filter using only single CCCDTA," Proceedings of the International Symposium on Communications and Information Technologies, vol. 8, pp. 305–309, 2008.
[26] J.W. Horng, C.L. Hou, C.M. Chang, Y.T. Lin, I.C. Shiu and W.Y. Chiu, "First-order allpass filter and sinusoidal oscillators using DDCCs," International Journal of Electronics, vol. 93, no. 7, pp. 457–466, 2006.
[27] M. A. Ibrahim, S. Minaei, and H. Kuntman, "DVCC based differential-mode all-pass and notch filters with high CMRR," International Journal of Electronics, vol. 93, pp. 231-240, 2006.
[28] B. Metin and O. Cicekoglu, "Novel cascadable allpass filter with a grounded capacitor," Proceedings of the 49th IEEE International Midwest Symposium on Circuits and Systems, pp. 242–244, 2006.
[29] K. Pal and S. Rana, "Some new first-order all-pass realizations using CCII," Active and Passive Electronic Components, vol. 27, pp. 91-94, 2004.
[30] M. A. Ibrahim, H. Kuntman, and O. Cicekoglu, "First-order all-pass filter canonical in the number of resistors and capacitors employing a single DDCC," Circuits Systems and Signal Processing, vol. 22, pp. 525-536, 2003.
[31] O. Çiçekoğlu, H. Kuntman, and S. Berk, "All-pass filters using a single current conveyor," International Journal of Electronics, vol. 86, pp. 947-955, 1999.S. Siripongdee and W. Jaikla, "Universal filter using single commercially available IC: LT1228," Proceedings of the 3rd International Conference on Mechatronics and Mechanical Engineering, 2016.

Acknowledgments
This research is financially supported by Faculty of Industrial Education and Technology, King Mongkut's Institute of Technology Ladkrabang (KMITL), Grant number: 2562-0203-009.

IOP Conf. Series: Materials Science and Engineering **559** (2019) 012010 doi:10.1088/1757-899X/559/1/012010

Realization of Voltage-mode Multifunction Biaquadratic Filter Using Minimum Number of Active Element

Winai Jaikla, Surapong Siripongdee and Peerawut Suwanjan

Department of Engineering Education, Faculty of Industrial Education and Technology, King Mongkut's Institute of Technology Ladkrabang, Bangkok, Thailand, 10520

winai.ja@kmitl.ac.th

Abstract. The realization of voltage-mode multifunction biquadratic filter has been presented in this research contribution. The proposed circuit contains three input voltage nodes and one output voltage node. The structure of proposed filter is simple consisting of one modified current controlled current conveyor transconductance amplifier (M-CCCCTA) as active building block with one resistor and two capacitors. With electronically controllable property of M-CCCCTA, the parameters, natural frequency (ω_0) and quality factor (Q) are electronically controlled by changing the external DC bias currents of the M-CCCCTA. The tuning characteristic of Q is also done without effect the w0. In the same circuit construction, five second order filter responses, all-pass, high-pass (HP), band-reject (BR), low-pass (LP) and band-pass (BP) functions are obtained. The output filter responses are obtained with applying appropriate input voltage to input node without the matching condition of active and passive elements. The performances of the proposed biquadratic filter have been verified through Pspice simulation in TSMC 0.25μm CMOS technology parameters.

1. Introduction

The realization of second order multifunction active filters which contain several filter responses such as band-pass, low-pass and high-pass functions in the same circuit topology receives high attention in analog signal processing system. Generally, the filter has been important in many fields for example in communication, instrument and sound systems etc. [1-2]. The active filter which uses the electronically controllable active building block can be easily adjusted the filter parameters such as the cut-off frequency, the quality facer, bandwidth and gain with microcontroller or microcomputer. This control is important in modern electronic systems [3].

The controlled current conveyor transconductance amplifier (CCCCTA) is a versatile active building block [4]. The CCCCTA can be employed to realize both current and voltage mode configurations. The transconductanc gain and its parasitic resistance are electronically tunable. With this characteristic, the CCCCTA based circuits can be easily controlled by microcomputer. A lot of electronic circuits employing CCCCTA have been proposed in the open literature. These circuits include analog multiplier [4], sinusoidal oscillator [5-9], inductance simulator [4, 10, 11] and first order filter [12-15] etc.

The voltage-mode second order multifunction filters using CCCCTA have been found in [16-24]. These filters include multiple inputs single output configuration [16, 19] and single input multiple outputs configuration [17, 18, 20, 21, 22, 23, 24]. However, these filters provide following drawbacks:

Content from this work may be used under the terms of the Creative Commons Attribution 3.0 licence. Any further distribution of this work must maintain attribution to the author(s) and the title of the work, journal citation and DOI.
Published under licence by IOP Publishing Ltd

- Require more than one active building block [17, 19, 20, 22, 23, 24].
- Require CCCCTA with multiple current output terminals [17, 18, 20, 21, 22].
- Require the matching condition to obtain output filter response [18, 20, 22].
- The control of parameter Q is not done without affecting the parameter ω_0 [16].
- Five filter responses are not obtained in the same circuit [18, 21, 23, 24].

This paper presents the design and analysis of the biquadratic filter emphasizing on the use of single active building block. The proposed filter is sort of multiple inputs and single output configuration with three inputs voltage and one output voltage. The selection of output voltage filter functions is done without the requirement of the critical matching condition of passive and passive elements. The quality factor of the proposed multifunction filer is tuned without effect the natural frequency.

2. Realization of Biquad Filter

The specifications of the M-CCCCTA used in this design, proposed filter, performance circuit analysis will be given in this section.

2.1. Active building block

The conventional CCCCTA [4] is the four terminals active building block. The two input terminals are y and x. The impedance at y terminal is high, while the resistance at x terminal (R_x) is controllability. The high impedance output terminals are z and o. The transconductance at o terminal (g_m) is also controllable. Generally, the R_x and g_m of CCCCTA are electronically controlled. To extend the use of CCCCTA, in this design, the high impedance voltage terminal (v) is added to the operational transconductance amplifier section and it will be called as modified current controlled current conveyor transconductance amplifier (M-CCCCTA). Figure 1 shows the circuit symbol of M-CCCCTA. While the equivalent circuit of M-CCCCTA is illustrated in Fig. 2. The terminal relationship of current/voltage in M-CCCCTA is defined by

$$I_y = 0;\ V_x = R_x I_x + V_y;\ I_z = I_x;\ I_v = 0;\ I_o = g_m\left(V_v - V_z\right) \tag{1}$$

In this design, the M-CCCCTA is implemented from CMOS transistors as drawn in Fig. 3. The parasitic resistance, R_x and transconductanc, g_m are given by

$$R_x = \frac{1}{\sqrt{8\mu_n C_{ox}(W/L)I_{B1}}} \text{ and } g_m = \sqrt{\mu_n C_{ox}(W/L)I_{B2}} \tag{2}$$

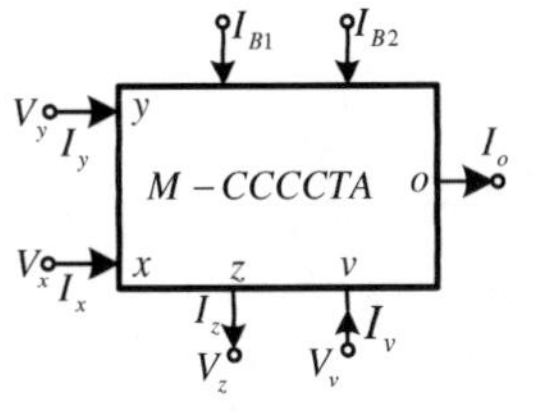

Figure 1. Symbol of M-CCCCTA

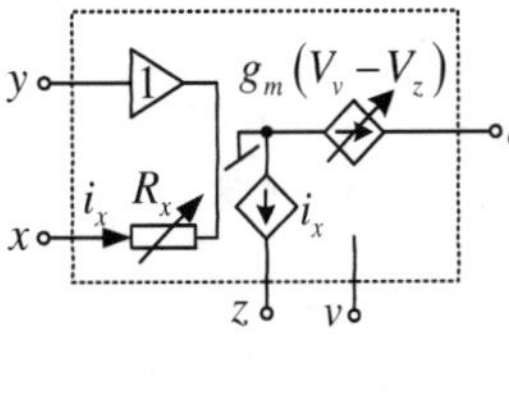

Figure 2. Equivalent circuit of M-CCCCTA.

2.2. Proposed Second Order Voltage-Mode Multifunction Filter

In Fig. 4, the proposed biquad multifunction filter with single M-CCCCTA is shown. Only three passive elements are required. This includes two capacitors C_1 and C_2, single resistor R. The proposed filter contains three input voltages, v_{i1}, v_{i2} and v_{i3} with single output voltage v_o. The proposed biquadratic filter is designed from the basic parallel RLC circuit where the M-CCCCTA and C_1 operate as inductance simulator. Using M-CCCCTA properties as described in section 2.1 and routine circuit analysis, the following output voltage is obtained

$$v_o = \frac{s^2 v_{i1} + s\frac{1}{C_2 R} v_{i2} + \frac{g_m}{R_x C_1 C_2} v_{i3}}{s^2 + s\frac{1}{C_2 R} + \frac{g_m}{R_x C_1 C_2}} \tag{3}$$

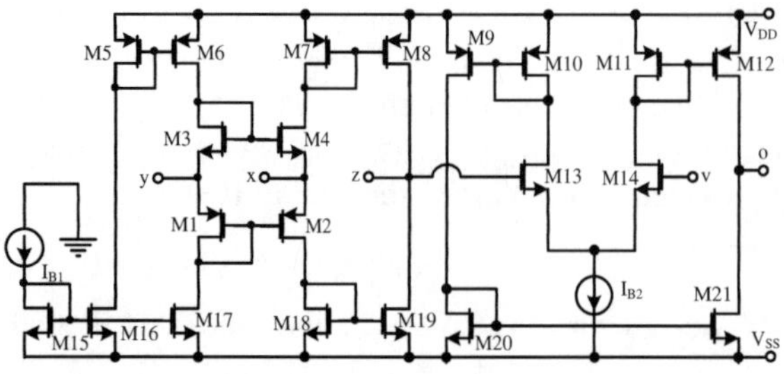

Figure 3. Internal construction of M-CCCCTA

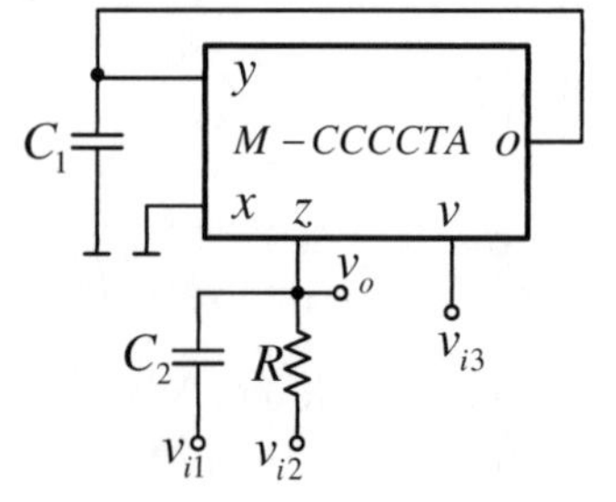

Figure 4. Proposed filter.

It should be noted form (3) that the different types of biquadratic filter can be obtained as follows:

- The biquadratic high-pass filter (HP) is realized by $v_{i1}=v_{in}$ and $v_{i2}=v_{i3}=0$.
- The biquadratic band-pass filter (BP) is realized by $v_{i2}=v_{in}$ and $v_{i1}=v_{i3}=0$.
- The biquadratic low-pass filter (LP) is realized by $v_{i3}=v_{in}$ and $v_{i1}=v_{i2}=0$.
- The biquadratic band-reject filter (BR) is realized by $v_{i1}= v_{i3}=v_{in}$ and $v_{i2}=0$.
- The biquadratic all-pass filter (AP) is realized by $v_{i1}=-v_{i2}= v_{i3}=v_{in}$.

It is found from (3) that all standard filter responses are obtained from the proposed biquadratic filter with same circuit configuration. Also, it is appeared that the voltage gain for each filter response is unit. The parameters ω_0 and Q are given by

$$\omega_0 = \sqrt{\frac{g_m}{R_x C_1 C_2}} \text{ and } Q = R\sqrt{\frac{g_m C_2}{R_x C_1}} \tag{4}$$

It is found from (4) that the parameter Q can be controlled by resistor R without effect the ω_0. Also ω_0 and Q can be electronically adjusted by mean of R_x or g_m by adjusting the auxiliary DC bias current I_{B1} or I_{B2}.

3. Simulation Results

Based on the 0.25μs TSMC CMOS process model (Level 3) [25], the presented three-inputs single-output multifunction second order filter in Fig. 4 was performed via PSpice program. The internal construction of CMOS M-CCCCTA in Fig. 3 was constructed. The symmetrical voltage supplies applied to the circuit were $V_{DD}=-V_{SS}=1.25$V. The channel width and length (W/L) of the NMOS and PMOS transistors are listed in Table 1. The external DC bias currents I_{B1} employed to control Rx and IB2 used to control gm were set to 80μA and 55μA, respectively. The value of resistor R was chosen as 1kΩ and capacitors C_1 and C_2 were chosen as 47pF. In Fig. 5, the simulated frequency response of voltage gain for low-pass, high-pass and band-pass functions is shown. It is found that the simulated natural frequency obtained from the proposed filter is 3MHz. For the band-reject and all-pass responses, the gain and phase responses are respectively shown in Fig. 6 and 7. It is clear from the results in Fig. 5 to Fig. 7 that the filter structure can completely provide five second order filter responses in the same configuration. By changing the values of resistor (R) to 1kΩ, 3kΩ and 10kΩ while other active and passive elements were same as mentioned above, the quality factors are respectively obtained as 0.63, 1.12 and 5.61 as appeared in Fig. 8. It is clearly that the tuning characteristic of parameter Q is controlled by resistor without distorting the parameter ω_0 as analyzed in Eq. (4). The time domain response of band-pass function is shown in Fig. 9 where the sinusoidal waveform with 100mVp-p and f=3MHz was fed to be the input voltage signal. The total harmonic distortion (THD) of the sinusoidal output voltage signal is 0.573%.

Table 1. Aspect ratio of MOS transistors

Transistor	W/L (μm)
M1, M2, M13, M14	16/0.5
M3 – M4, M5 – M12	10/0.5
M15 – M21	6/0.5

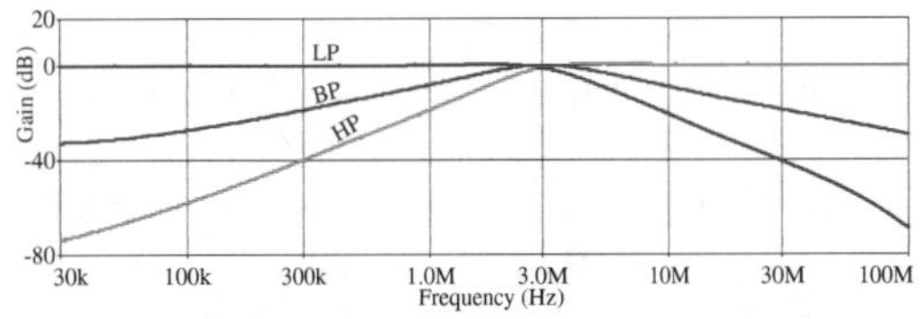

Figure 5. Simulated amplitude frequency responses of high-pass, band-pass, and low-pass functions.

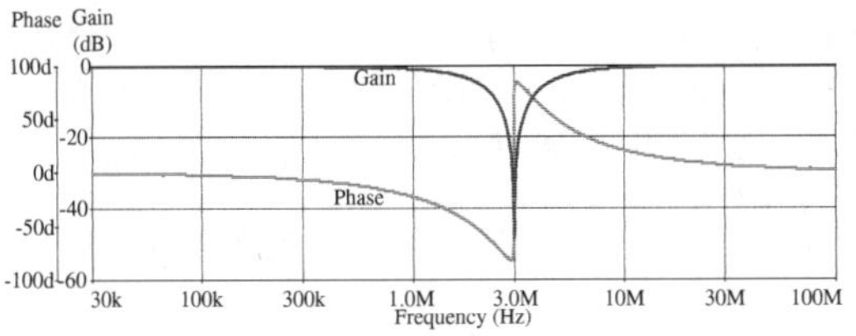

Figure 6. Simulated amplitude and phase frequency response of band-reject function..

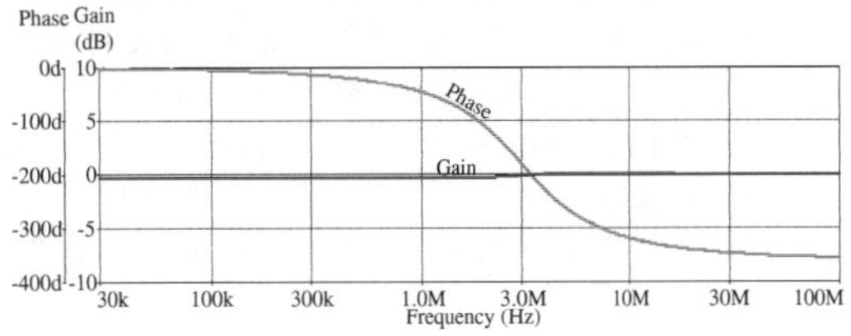

Figure 7. Output spectrum.

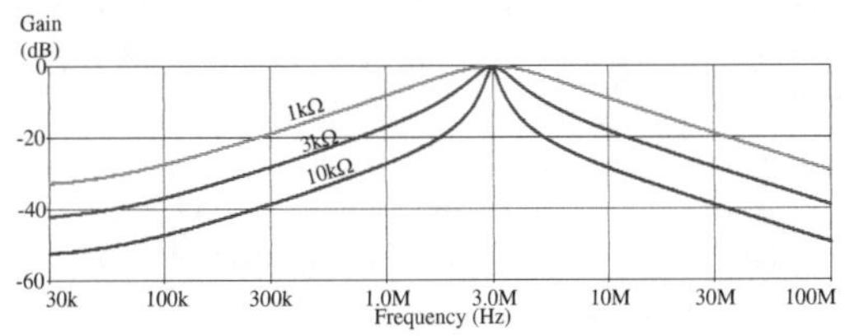

Figure 8. Simulated gain and phase response

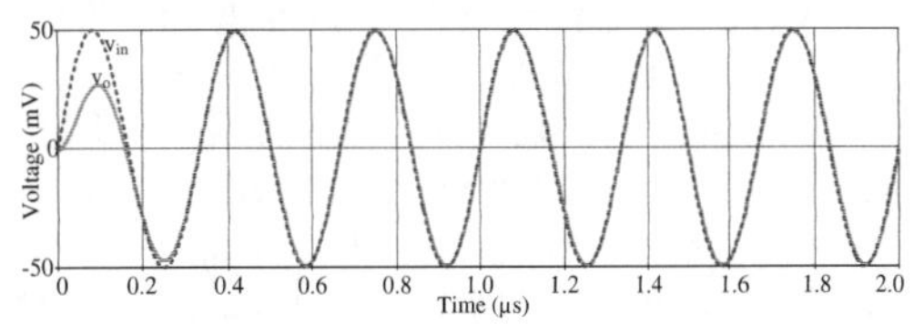

Figure 9. Transient response of BP function.

4. Conclusion

The active filter is design in this paper. The proposed multifunction filter consists of on M-CCCCTA with one resistor and two capacitors. It has three input voltages and one output voltage. It can provide five voltage transfer functions of second order filter. The selection of the output filter responses can be performed without any matching condition requirement. The tuning property of ω_0 and Q is electronically done. The performances of the proposed biquadratic filter were performed through Pspice program in 0.25μs TSMC CMOS process. Simulation results for all filter functions agree well as theoretical analysis. Moreover, the control of quality factor is achieved without disturbing the natural frequency as analyzed in (4) and verified via simulation result in Fig. 8.

5. References

[1] R. Verma, N. Pandey, and R. Pandey, "Electronically Tunable Fractional Order Filter," Arabian Journal for Science and Engineering, vol. 42, pp. 3409-3422, Aug 2017.

[2] P. Supavarasuwat, M. Kumngern, S. Sangyaem, W. Jaikla, and F. Khateb, "Cascadable independently and electronically tunable voltage-mode universal filter with grounded passive components," AEU - International Journal of Electronics and Communications, vol. 84, pp. 290-299, 2018.

[3] C. Psychalinos, C. Kasimis, and F. Khateb, "Multiple-input single-output universal biquad filter using single output operational transconductance amplifiers," AEU - International Journal of Electronics and Communications, vol. 93, pp. 360-367, 2018.

[4] M. Siripruchyanun and W. Jaikla, "Current controlled current conveyor transconductance amplifier (CCCCTA): a building block for analog signal processing," Electrical Engineering, vol. 90, pp. 443-453, Jun 2008.

[5] Y. A. Li, "Further Research on Systematic Synthesis for CCCCTA Quadrature Oscillators," Recent Advances in Electrical & Electronic Engineering, vol. 10, pp. 209-215, 2017.
[6] H. P. Chen, Y. S. Hwang, and Y. T. Ku, "Voltage-mode and current-mode resistorless third-order quadrature oscillator," Applied Sciences (Switzerland), vol. 7, 2017.
[7] Y. A. Li, "Systematic Derivation for Quadrature Oscillators Using CCCCTAs," Radioengineering, vol. 24, pp. 535-543, Jun 2015.
[8] H. P. Chen, S. F. Wang, and Y. T. Ku, "CCCCTA-based resistorless voltage and current mode quadrature oscillator," IEICE Electronics Express, vol. 12, Jul 10 2015.
[9] Y. A. Li, "Derivation for current-mode Wien oscillators using CCCCTAs," Analog Integrated Circuits and Signal Processing, vol. 84, pp. 479-490, Sep 2015.
[10] S. Lawanwisut and M. Siripruchyanun, "An electronically controllable active-only current-mode floating inductance simulator," Proceedings of the 35th International Conference on Telecommunications and Signal Processing, Prague, 2012, pp. 386-389.
[11] N. Pandey, R. Bazaz, and R. Manocha, "MO-CCCCTA-based floating positive and negative inductors and their applications," Journal of Electrical and Computer Engineering, Article ID 150354, 2011.
[12] K. Chinpark, W. Jaikla, S. Siripongdee, and P. Suwanjan, "Electronically controllable first-order multifunction filter with using single active building block," Proceedings of the 3rd International Conference on Control and Robotics Engineering (ICCRE), 2018, pp. 192-195.
[13] M. Kumngern and J. Chanwutitum, "An electronically tunable current-mode first-order allpass filter using a CCCCTA," Proceedings of the 2013 International Conference on Advanced Technologies for Communications, 2013, pp. 733-736.
[14] W. Jaikla, A. Noppakarn, and S. Lawanwisut, "New gain controllable resistor-less current-mode first order allpass filter and its application," Radioengineering, vol. 21, pp. 312-316, Apr 2012.
[15] N. Pandey, P. Arora, S. Kapur, and S. Malhotra, "First order voltage mode MO-CCCCTA based all pass filter," Proceedings of the 2011 International Conference on Communications and Signal Processing, Kerala, 2011, pp. 535-537.
[16] N. Gupta and P. P. Bansod, "Programmable voltage mode multifunctional filter using modified CMOS current controlled current conveyor transconductance amplifier," Proceedings of the 2nd International Conference on Communication, Control and Intelligent Systems, 2016, pp. 185-190.
[17] H. P. Chen and W. S. Yang, "Electronically tunable current controlled current conveyor transconductance amplifier-based mixed-mode biquadratic filter with resistorless and grounded capacitors," Applied Sciences (Switzerland), vol. 7, Mar 2017.
[18] S. V. Singh, S. Maheshwari, R. S. Tomar, and D. S. Chauhan, "Single input four output voltage-mode biquad filter with electronic tuning," Proceedings of the 2013 International Conference on Multimedia Signal Processing and Communication Technologies, Aligarh, 2013, pp. 226-230
[19] P. Beg, R. S. Badoni, M. S. Ansari, and P. Kulshreshtha, "Programmable voltage mode resistorless multifunctional filter using CCCCTA," Proceedings of the 4th International Conference on the Next Generation Information Technology Summit, Noida, 2013, pp. 441-445.
[20] S. V. Singh, S. Maheshwari, and D. S. Chauhan, "Novel electronically tunable mixed-mode biquad filter," Proceedings of the 2011 International Conference on Electric and Electronics, Nanchang, 2011, pp. 735-742.
[21] S. V. Singh, S. Maheshwari, and D. S. Chauhan, "Single CCCCTA-based SITO biquad filter with electronic tuning," Proceedings of the 2011 International Conference on Multimedia, Signal Processing and Communication Technologies, Aligarh, 2011, pp. 172-175.
[22] S. Maheshwari, S. V. Singh, and D. S. Chauhan, "Electronically tunable low-voltage mixed-mode universal biquad filter," IET Circuits Devices & Systems, vol. 5, pp. 149-158, May 2011.
[23] W. Jaikla, A. Lahiri, A. Kwawsibsame, and M. Siripruchyanun, "High-input impedance voltage-mode universal filter using CCCCTAs," Proceedings of the 7th Annual International Conference on Electrical Engineering/Electronics, Computer, Telecommunications and Information Technology, ECTI-CON 2010, Chiang Mai, 2010, pp. 754-758.

[24] S. V. Singh, S. Maheshwari, J. Mohan, and D. S. Chauhan, "An electronically tunable SIMO biquad filter using CCCCTA," Communications in Computer and Information Science, vol. 40, 2009, pp. 544-555.
[25] P. Prommee and M. Somdunyakanok, "CMOS-based current-controlled DDCC and its applications to capacitance multiplier and universal filter," AEU-International Journal of Electronics and Communications, vol. 65, pp. 1-8, 2011.

Acknowledgments
This research is financially supported by Faculty of Industrial Education and Technology, King Mongkut's Institute of Technology Ladkrabang (KMITL).

IOP Conf. Series: Materials Science and Engineering **559** (2019) 012011 doi:10.1088/1757-899X/559/1/012011

Voltage Supply Assessment Using Process Capability Concepts: A Case Study

A H Vizuete[1], G J Barbosa[1], E M Lema[1], M S Albán[1], R S Vargas[2] and L C Ruiz[3]

[1]Faculty of Applied Sciences and Engineering, Universidad Técninca de Cotopaxi, 050161 Latacunga, Simon Rodriguez Av., Ecuador.
[2]Doctoral School on Material Sciences and Technologies, Obuda University, 1034 Budapest, Beci 94-96, Hungary.
[3]Doctoral School on Safety and Security Sciences, Obuda University, 1081 Budapest, Népszínház 8, Hungary.

vargas.ramiro@phd.uni-obuda.hu

Abstract. The following case study examines a company which experienced several difficulties caused by low-quality voltage supply. Two main points were considered in the analysis, one located next to the substation transformer and a second one inside the company. Voltage parameters such as harmonics presence, voltage flickers, average voltage, and power factor were measured in this study. As a result, 62 000 samples were collected. The data obtained from point 2 (control group) revealed the necessity of introducing voltage regulators in order to improve the energy efficiency and save in energy costs at the company. Therefore, 1 500 samples were collected at point 2 (treatment group) to verify the voltage regulators' efficiency. Comparing the control and treatment group, there was a significant improvement of 93% regarding process capability in power factor phase CN.

1. Introduction

Energy sector faces serious problems nowadays [1]. Voltage supply quality is the main concern regarding work efficiency in any industry. The maintenance processes for controlling voltage stability generate uncertainty towards electric supply-demand [2]. The intense load in transmission lines causes voltage related problems in different industries [3]. Furthermore, automated industrial processes induce the incorporation of nonlinear loads, which are important perturbation sources in electric systems [4]. These disturbances inject current harmonic that distorts the fundamental waveform. Hence, when an interaction with the system happens, it distorts the voltage and provokes problems in the equipment that comprises the system.

In Ecuador, the rising development of the industries has resulted in a more demanding market for energy suppliers. Not only in terms of the service but also, requiring better standards for the energy delivered [5]. Legal regulations are administered in Ecuador by the regime law of the electric sector. These standards establish responsibilities between the consumer and the suppliers. National and international norms limit the permitted ranges of voltage waveform distributed by the supplier and distortions made by the consumers in their production processes.

At an international level, the regulation that governs these types of systems is IEEE 5191992. This norm includes the harmonic distortions introduced by nonlinear loads in order to solve problems regarding potency quality [6]. The norm IEC 61000-4-30 is widely used for measuring processes of

Content from this work may be used under the terms of the Creative Commons Attribution 3.0 licence. Any further distribution of this work must maintain attribution to the author(s) and the title of the work, journal citation and DOI.
Published under licence by IOP Publishing Ltd

electricity quality [7]. Additionally, in Ecuador, the CONELEC 004/01 norm establishes a regulating format of the electricity quality distribution, which allows locating in an efficient way the quality in three stages: product, technical service and commercial service [8]. This regulation points out the energy quality evaluation as a form to improve its efficiency. Moreover, it determines corrective actions in order to suppress the perturbation content in the main network feeder.

Consequently, the following case study describes the evaluation of the voltage source at a wood-processing company located in Ecuador with almost 40 years of experience in the market. In 2000 and 2006, the industry went through an evaluation of national institutes which suggested few changes to improve the energy consumption. Some of the suggestions included motors and light bulbs replacement and the installation of new instruments to regulate the voltage consumption, especially during rush-hours.

The main objective of this study is to determine a strategy to reduce the most relevant perturbations regarding the quality of the electric supply. Utilizing as a reference, national and international norms that regulate the limits, measurement periods and compliance stages.

2. Methodology

A preliminary assessment was performed to evaluate the voltage efficiency using two points of reference:

- The first point was located at the distribution transformer which provides energy to the whole town.
- The second point was located inside the company studied at 800 meters distance from the first point.

As the main feeder delivers line voltage and phase voltage, to have a point of reference, only phase voltages were studied. Thus, voltage phases' parameters were treated separately as AN, BN and CN.

Table 1 enumerates the time taken to collect the average voltage, harmonics presence, electrical flickers, and power factor samples. Moreover, the electrical network configuration, nominal voltage, working frequency, and the equipment used to measure the data.

Table 1. Initial analysis - points' details.

	Point 1	**Point 2**
Analysis time length	62 days	14days
Configuration	3 Phases + Neutral Wye configuration	
Phase voltage	13800 V	440 V
Frequency	60 Hz	
Analysis instrument	Fluke 1750	

As a result of this preliminary analysis, three Siemens JFR Distribution Step-Voltage Regulators in delta configuration were installed in the company. A second assessment of 10 days in point 2 was performed in order to test the results of implementation of the voltage regulators. Point one measurement was analyzed to obtain baseline information regarding the energy quality delivered by the local electricity provider.

CONELEC 004/01 norm envisages voltage as a product delivered by national electric power industries. Thus, a process capability analysis was performed on the data sets collected. According to CONELEC 004/01, the product has to fulfill specific requirements regarding three main aspects: average voltage, voltage flicker and power factor. In the case of voltage flicker, the norm considers only short-term flicker perceptibility (Pst) and all data has to be collected under the International Electrotechnical Commission (IEC) norm 60868. Since process capability indices can be understood as the number of non-conforming parts per million [9], this concept accords with CONELEC 004/01 assessment parameters.

Data sets were analyzed using the statistical package Minitab. Process Capability Sixpack report was utilized to compare the intervention effects of the voltage regulators. Point 2 of the initial analysis is

IOP Conf. Series: Materials Science and Engineering **559** (2019) 012011 doi:10.1088/1757-899X/559/1/012011

considered as the control group while the data collected at the secondary assessment is the treatment group.

3. Results

This section presents the data collection of three factors: power factor, voltage flickers and the average voltage at point 2. Harmonic distortion was not considered for the results since it did not alter the voltage efficiency. In the preliminary assessment, 62 000 samples were taken, whereas over 1 500 samples were taken after the installation in the second point. However, only the treatment group data is in detail described for each voltage distribution phase.

3.1. Power Factor

According [10] a process capability lesser than 0.5 means at least 13.36% of the data is out of tolerance Figure 1, Figure 2 and **Figure 3** show that process capability (cp) at the phase CN is out of tolerable limits after the implementation of the step voltage regulator. Figure 4 shows a slight improvement compared with the values collected in the preliminary assessment. Hence, AN phase and BN phase meet acceptable quality parameters.

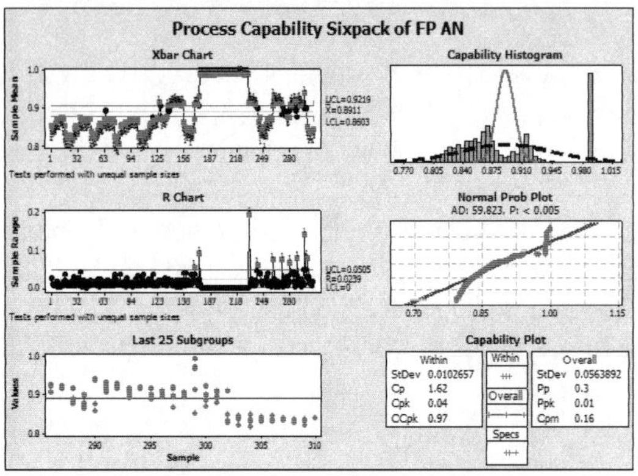

Figure 1. Average Power Factor AN.

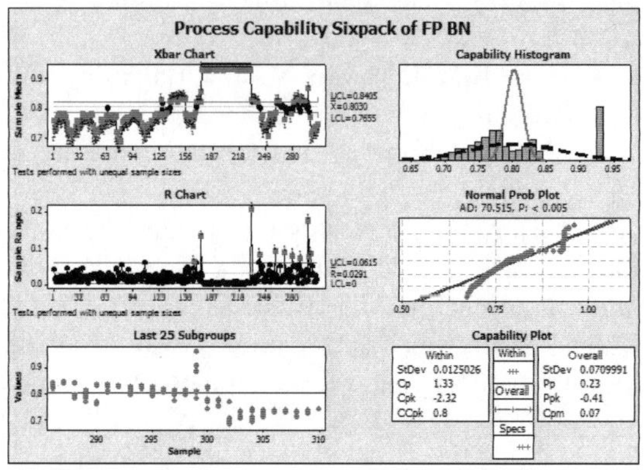

Figure 2. Average Power Factor BN.

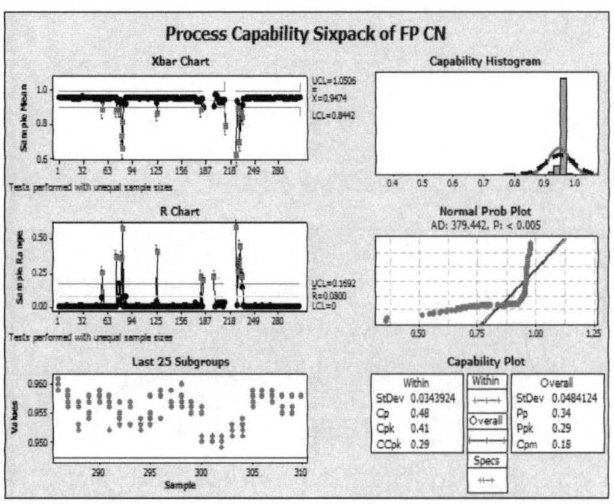

Figure 3. Average Power Factor CN.

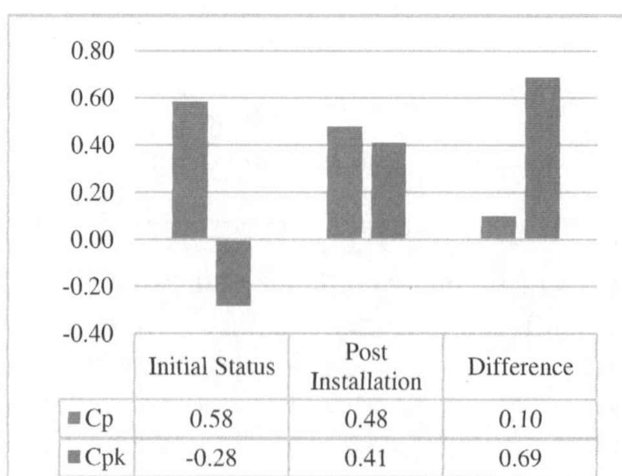

Figure 4. Cp and Cpk at point 2. CN power factor's changes analysis.

Figure **4** shows the initial state of the process and the effect of the treatment in process capability (Cp, CpK) The increment of 0.69 in Cpk translates to a considerable improvement of 93%.

3.2. Power - line flickers

Figure 5, Figure 6 and **Figure 7** summarizes short term flicker perceptibility (Pst) of AN, BN and CN. The IEC61000-4-15 flickermeter standard establishes that Pst should ideally be 1.00 with a tolerance of 5%. An Upper Control Limit of 1.35, 1.30 and 1.34 from AN. BN and CN Xbar chart respectively, demonstrates that all phases are out of limits regarding Pst.

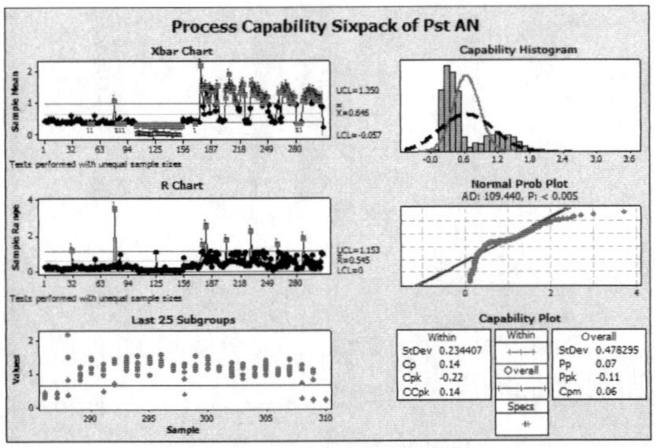

Figure 5. Pst - Phase AN.

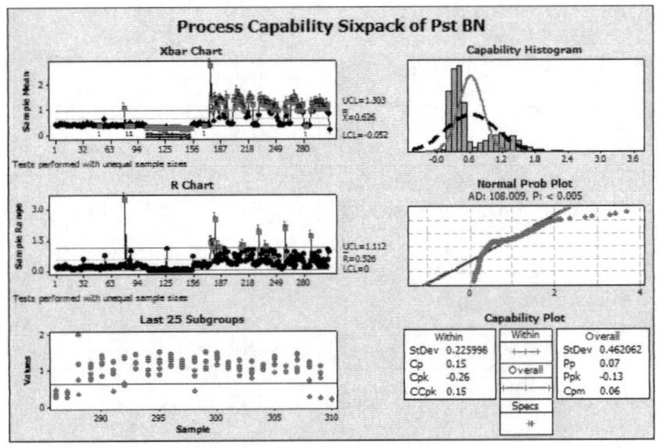

Figure 6. Pst - Phase BN.

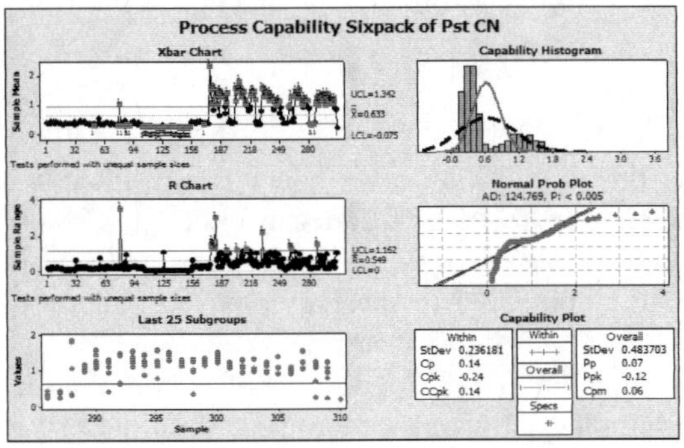

Figure 7. Pst - Phase CN.

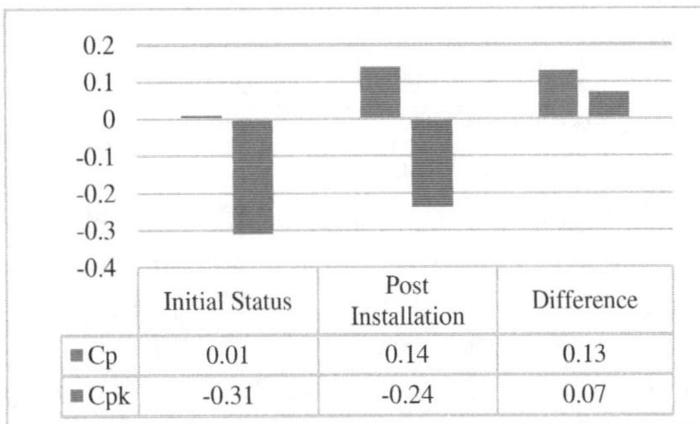

	Initial Status	Post Installation	Difference
Cp	0.01	0.14	0.13
Cpk	-0.31	-0.24	0.07

Figure 8. Cp and Cpk at point 2. CN Pst's changes analysis.

Nonetheless, **Figure 8** shows there was a slight improvement in comparison with the initial state at phase CN. This, due to the fact that external loads cause voltage flickers to the company.

3.3. Average voltage

Figure 9, Figure 10 and **Figure 11** show a satisfying cp in all cases AN, BN and CN average voltage. With a cp of 19.98, 18.72 and 20.1 at AN, BN and CN respectively, the process achieved a six-sigma performance. As reported by [11], a six-sigma process does not produce more than 4 defects per million opportunities. Since the main objective of a step voltage regulator is to stabilize nominal voltage, then the instrument is completely efficient.

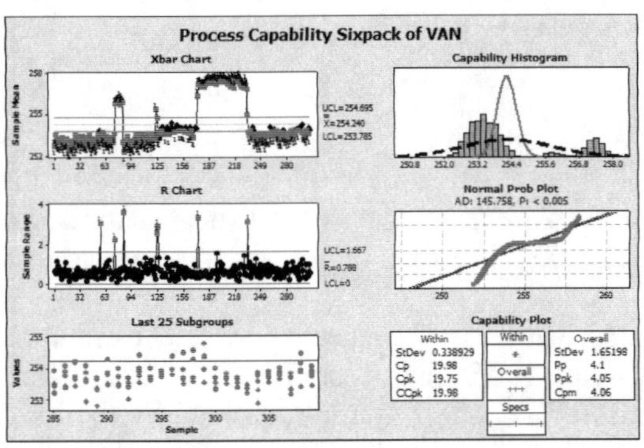

Figure 9. Average Phase Voltage AN.

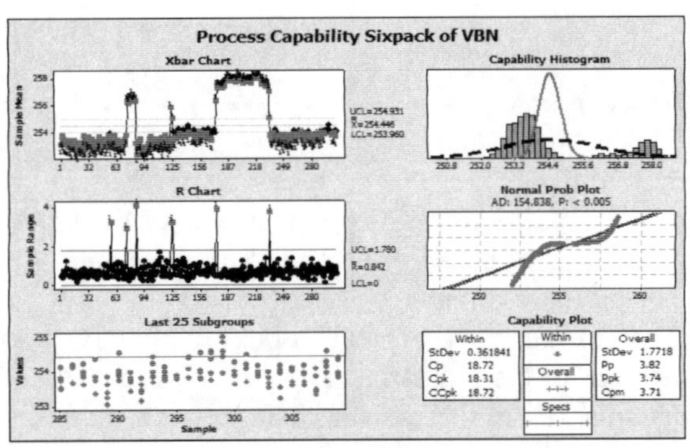

Figure 10. Average Phase Voltage BN.

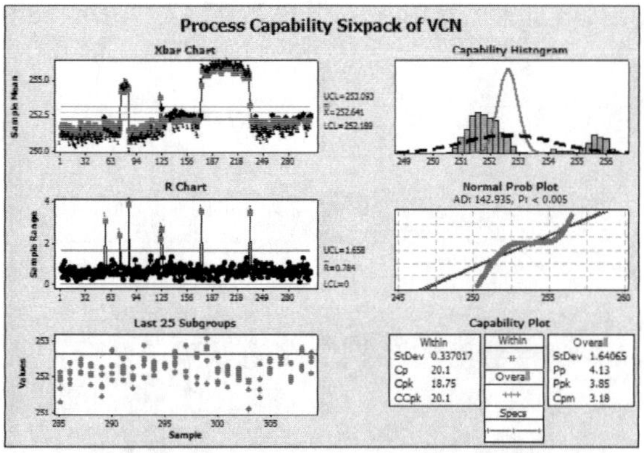

Figure 11. Average Phase Voltage CN.

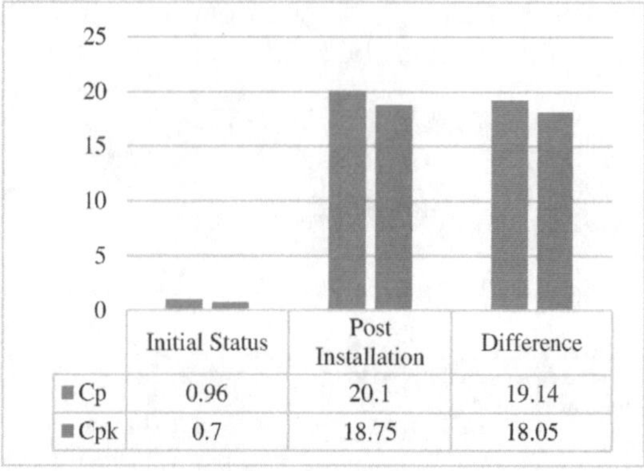

Figure 12. Cp and Cpk at point 2. CN average voltage changes analysis.

However, Figure 12 highlights the most significant improvement from all the voltage parameters studied.

3.4. Economic analysis

In 2014, the company experienced working hours loses due to multiple power outages. It resulted in a financial loss of over 190000 USD. Therefore, voltage step regulators and capacitor banks were implemented. Figure 13 shows the economic losses in 2015. As a result, the losses were reduced in nearly 93% which means 175000 USD in the year 2015 compared to the previous year where no treatment was applied.

Almost 100% of the voltage variations of the samples are within the allowed limits of ± 8.0% that the regulation specifies. In relation to flicker, 24 % of the samples per phase were outside the limit that is 1, and with respect to power factor, 54% of the samples were less than the limit of 0.92. In the same way, the presence of harmonics was verified but they did not exceed the allowed limits.

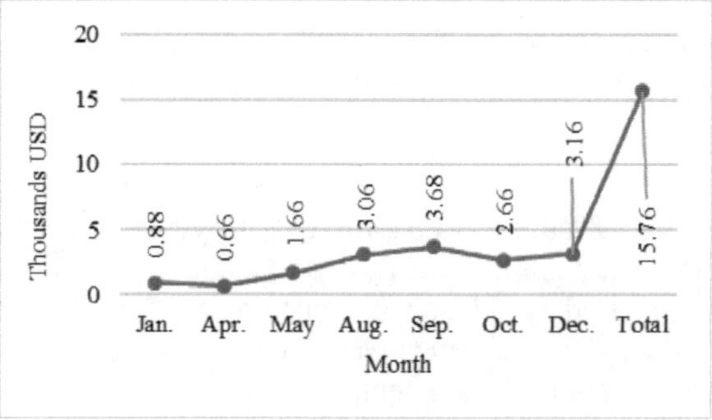

Figure 13. Losses due to power outage – year 2015.

4. Discussion

The analysis at the delivery point (Point 1, main feeder) and the output of the transformer (Point 2) showed that parameters such as: voltage, harmonics and flicker were outside the limits allowed by the national electricity regulation institute (CONELEC 004/01). After technical and statistical analysis, the installation of banks of capacitors and regulators was justified. The final effect was the correction of most of these parameters except flicker.

The infringement of regulations are mainly associated with users connected to the distribution network in the industrial range. Electric arc furnaces and high energy consumption loads introduce disturbances in the transmission network. For this reason, the energy quality provided by the local energy supplier at the delivery point of analysis (Point 1) is affecting the electric distribution in the system.

On the other hand, current harmonics, as well as the unbalance in the 13.8 kV bus of the Local Feeder No. 5 substation, were significant. However, one of the functions of the voltage regulator is that it works as a harmonic filter. Therefore it is possible to reduce them and satisfy the requirements of the Energy Quality Regulator.

The power factor at the delivery point (Point 1) is outside the permitted values according to the regulation. Thus, a 300 kVA capacitor bank was connected in parallel to the load in order to correct this requirement. It can be verified that the voltage variation has been controlled with the installation of the single-phase voltage regulators, and it also shows that they serve as low-pass filters, decreasing the harmonic content.

It is observed that the process capability indexes Cp and Cpk increased after the installation of the voltage regulators per face and capacitor banks. This showed that the variability of the parameters of voltage, flicker and power factor were optimized as a result of the corrections made. Additionally, during the year 2015, the company did not invest in replacing broken electrical equipment due to the voltage disturbances. Certainly, it cannot be adduced that the step voltage regulator completely resulted in the financial improvement. Some electrical equipment could have had their own failures due to the length of time they had been used.

5. Conclusions

Energy efficiency is a paramount concern within small, medium and large enterprises because of the economic savings it represents for a business. Ecuadorian electricity distribution causes economic losses for corporations. For this reason, companies are in the need to find solutions inside their production processes. The study presented above captured how small interventions can contribute to great improvements in productivity and reducing costs.

The case study evaluated electricity efficiency at a wood manufacturer company located in Ecuador. A preliminary analysis of voltage efficiency was performed measuring different electrical factors at two critical points: outside the company at the distribution transformer and the second point inside the company. As a result of this analysis, voltage regulators were set up at company's premises.

In order to verify the success of this interventions, further measures were taken at point 2 and compared with the previous samples collected at the preliminary study. The results obtained revealed a great improvement in the process capability of 93% regarding power factor at one of the voltage lines and an astonishing reduction in costs. This study represents a proof that simple informed decisions based on proper analysis can lead to considerable changes in productivity and competitive advantages that benefit an organization.

6. References

[1] G. Verbong and F. Geels, ""The ongoing energy transition: lessons from a socio-technical, multi-level analysis of the Dutch electricity system (1960–2004)," Energy policy, vol. 35, no. 2, pp. 1025-1037, 2007.

[2] D. Wang, S. Parkinson, W. Miao, H. Jia, C. Crawford and N. Djilali, "Online voltage security assessment considering comfort-constrained demand response control of distributed heat pump systems," Applied Energy, vol. 96, pp. 104-114, 2012.

[3] D. Devaraj and J. P. Roselyn, "On-line voltage stability assessment using radial basis function network model with reduced input features," International Journal of Electrical Power & Energy Systems, vol. 33, no. 9, pp. 1550-1555, 2011.

[4] M. M. Swamy, S. L. Rossiter, M. C. Spencer and M. Richardson, "Case studies on mitigating harmonics in ASD systems to meet IEEE 519-1992 standards," Industry Applications Society Annual Meeting, 1994., Conference Record of the 1994 IEEE, vol. 1, pp. 685-692, 1994.

[5] J. L. Espinoza and H. Vredenburg, "Towards a model of wind energy industry development in industrial and emerging economies," Global Business and Economics Review, vol. 12, no. 3, pp. 203-229, 2010.

[6] M. Halpin, "Overview of revisions to IEEE standard 519-1992," Quality and Security of Electric Power Delivery Systems, 2003. CIGRE/PES 2003. CIGRE/IEEE PES International Symposium, pp. 65-68, 2003.

[7] A. E. Legarreta, J. H. Figueroa and J. A. Bortolin, "An IEC 61000-4-30 class A—Power quality monitor: Development and performance analysis," Electrical Power Quality and Utilisation (EPQU), 2011 11th International Conference on, pp. 1-6, 2011.

[8] O. Castañeda and W. Castañeda, Análisis de Calidad de Energía acerca de la Calidad del Producto e Influencia de Armónicos de Corriente dentro del Área de Concesión de CNEL-Milagro, Corporación Centro Nacional de Control de Energía-CENACE Eléctrica de Guayaquil, 2010.

[9] V. E. Kane, "Process capability indices," Journal of quality technology, vol. 18, no. 1, pp. 41-52, 1986.

[10] D. C. Montgomery and G. C. Runger, Applied Statistics and Probability for Engineers, New York: John Wiley & Sons, Inc, 2003.

[11] T. Pyzdek and P. A. Keller, The six sigma handbook, vol. 4, New York: McGraw-Hill Education, 2014.

Effect of Calcination Temperature and Support Type of Pt/WO$_x$/boehmite Catalyst on 1,3-propanediol Production from Hydrogenolysis of Glycerol

P Uttraporn[1] and P Praserthdam[1,a]

[1] Center of Excellence on Catalysis and Catalytic Reaction Engineering, Department of Chemical Engineering, Faculty of Engineering, Chulalongkorn University, Phyathai Road,Patumwan, Bangkok, 10330, Thailand

[a] Corresponding author: piyasan.p@chula.ac.th

Abstract. Hydrogenolysis of glycerol to 1,3-PDO was tested over different catalysts. To enhance 1,3-PDO selectivity, our work raised the calcination temperature from 800 °C to 900 °C that promoted Bronsted acidity, glycerol conversion and 1,3-PDO selectivity under the mild reaction condition (140 °C, 0.5 MPa). Glycerol conversion gradually raised from 32.8% to 36.6% and 1,3-PDO selectivity mildly increased from 18.2% to 21.2%. The loading of SiO_2 into Pt/WO$_x$/γ-Al_2O_3 had extremely favourable effects on 1,3-PDO selectivity because the amount of Bronsted acid sites increased significantly as confirmed by NH_3-IR.

1. Introduction

Nowadays, the industrial manufacturing has a policy to promote the production of biodiesel from natural products in order to reduce the import of petroleum, in which the price is unpredictable and the loss of currency could be huge. The renewable biodiesel is an alternative energy, where its impact and significance have drawn tremendous interest both industrially and academically. Biodiesel can be made from various renewable sources such as plant oil, animal fat, waste cooking oil, etc.

Normally, biodiesel is acquired from the tranesterification method of triglyceride (fat and oil) in which 90% of methyl ester (biodiesel) and 10% of glycerol are created. Likewise, glycerol can be produced from saponification process and hydrolysis reaction. It is in concern that huge amounts of glycerol are synthesized in biodiesel reaction. The tranesterification process generated biodiesel and glycerol at a volumetric ratio of 10:1, so for every 1 m3 of biodiesel, 0.1 of crude glycerol is obtained [1].

Glycerol is a valuable by product because it can be used in several applications for the industry. In the past, glycerol has more than two thousand different applications, especially in pharmaceuticals, personal care, foods and cosmetics. Glycerol is a nontoxic, editable, biodegradable feedstock. So, it will support significant environmental benefits to the new principles products. Glycerol is mainly used in the process of drugs, medicine and pharmaceuticals for the purpose of dissolving drug, giving the pills humidity and enhancement the viscosity of liquid drugs [2]. Glycerine or glycerol is an outstanding solvent of iodine, bromine, phenol, tannins, alkaloids, and mercury chloride [3]. Accordingly, all applications are not enough to manage amounts of surplus glycerol in global. Thus, the new valuable product is desired to get rid of an excess glycerol.

Essentially, new applications or new value-added products from glycerol must have been created by various pathways of chemical reaction. In the past, there were so many conversion processes of

Content from this work may be used under the terms of the Creative Commons Attribution 3.0 licence. Any further distribution of this work must maintain attribution to the author(s) and the title of the work, journal citation and DOI.
Published under licence by IOP Publishing Ltd

glycerol that have been reported including oxidation, dehydration, esterification, acetylation, reforming, hydrogenolysis, etherification, ammoxidation, acetalization and emzymatic reactions [4].
The entire products from alteration of glycerol, 1,3-propanediol and 1,2-propanediol are main products of hydrogenolysis reaction from glycerol because these two chemicals are essential substances in the process of synthesizing products or goods in the market. 1,2-propanediol is harmless chemical that can be synthesized and used as antifreeze deicing agents, pharmaceuticals, food, cosmetics, liquid detergents, flavorings and personal care products [5], since 1,3-propanediol can be used for various applications, especially as an intermediate to produce polymethylene terephthalate (PTT), polyether and polyurethane [6]. From the molecular structure, glycerol has one more –OH group than propanediol, so the hydrogenolysis reaction of glycerol to propanediol contains of the removal of one H_2O and the addition of one H_2 molecule. The mechanism of hydrogenolysis reaction of glycerol can be divided into 3 types: dehydration-hydrogenation, dehydrogenation-dehydration-hydrogenation and direct glycerol hydrogenolysis [7]. Exactly, the mechanism should be specific, however it depends on the properties of catalyst used in reaction, such as the different acid-base catalysts, which can lead to different reaction mechanisms including the stability of intermediate in the reaction [8].
According to the literature, the hydrogenolysis catalysts have been applied in the hydrogenolysis reaction. It is found that there are two main functions to the selected catalyst for the hydrogenolysis reaction. First, the acidity or basicity of catalyst used to eliminate an OH-group. Mostly, metal oxides or acidic or basic support are used to provide acid-base function. The other one is the oxidation - reduction of hydrogen in the hydrogenolysis reaction. By the way, a lot of metals are also used as catalyst. It can be divided into 2 groups: transition metals (Cu, Ni, Co, etc.) [9-11] or noble metal (Pt, Ir, Ru, etc.) [12, 13]. Noble metals have remarkable properties in the oxidation-reduction reaction of hydrogen in the hydrogenolysis process. Kurosaka *et al.* reported that hydrogenolysis of glycerol reaction over $Pt/WO_3/ZrO_2$ catalyst in DMI gave 28% selectivity to 1,3-propanediol at 86% conversion [14]. Zhu *et al.* using $Pt\text{-}H_4SiW_{12}O_{40}/SiO_2$ catalyst in aqueous phase at 200 °C, 6 MPa, showed 31.4% yield of 1,3- propanediol and 64.5% glycerol conversion [15]. Tao *et al.* carried out mesoporous WO_3 supported Pt catalyst afforded 18.0% conversion of glycerol and 39.2% selectivity for 1,3-propanediol [15]. In 2010-2012, Shinmi and Tomishige reported the use of $Rh\text{-}ReO_x/SiO_2$ and $Ir\text{-}ReO_x/SiO_2$ as catalysts for hydrogenolysis reaction of glycerol in water. The condition of reaction is 120 °C and 8 MPa. The result showed high yield of 1,3- propanediol. The mechanism of reaction is direct glycerol hydrogenolysis (route 3, scheme 1). Glycerol is adsorbed on surface of ReO_x cluster at the terminal or middle position to generate 2,3-dihydroxypropoxide and 1,3-dihydroxypropoxide, respectively. Next, hydrogen activated on Ir metal attacks on 2-position of 2,3-dihydroxypropoxide to form 3-hydroxypropoxide. The hydrolysis of 3-hydroxypropoxide produces 1,3-propanediol. If hydrogen attacks on 3-position of 1,3-dihydroxypropoxide, the product will be 1,2- propanediol [16-18]. Quin *et al.* reported $Pt/WO_3/ZrO_2$ catalyst at 130 °C and 4 MPa and showed a high yield of 1,3-propanediol (32%) [19]. Gong *et al.* obtained 15.3% glycerol conversion and 50.5% 1,3-propanediol selectivity over $Pt/WO_3/TiO_2/SiO_2$ catalyst. [20]. In 2014, Kaneda *et al.* reported using Pt/WO_3/''AlOOH'' catalyst to obtain the highest 66% 1,3-propanediol because of the abundant of Al-OH groups in boehmite support [21].
From previous work, the hydrogenolysis reaction of glycerol was carried out at high pressure and temperature. So, our study was performed in mild condition and varied different supports by adding SiO_2 and different calcination temperatures of "AlOOH" (800 °C and 900 °C).

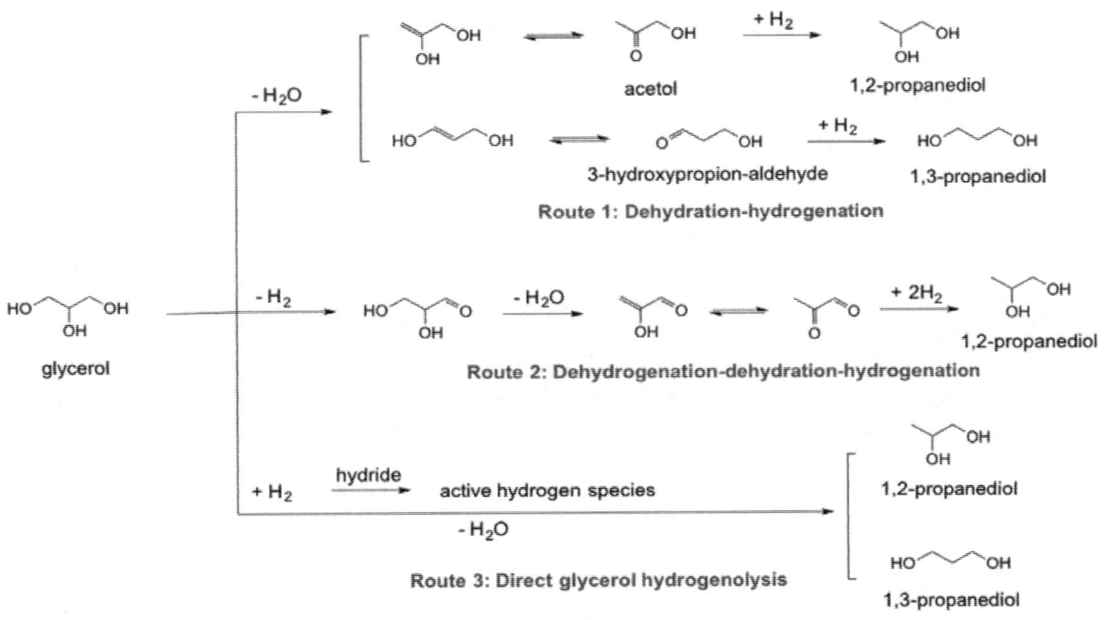

Sheme 1. The pathways of glycerol to propaindiol by hydrogenolpysis [22].

2. Experimental

2.1. Chemicals

1,3-propanediol (1,3-PDO), 1,2-propanediol (1,2-PDO) were purchased from Wako Pure Chemicals Co. Chloroplatinic acid (H_2PtCl_6), glycerol, ammonium (Meta) tungstate were purchased from Sigma-Aldrich, colloidal silica and boehmite (AlOOH) were obtained from the Taimei Chemicals Co. colloidal silica.

2.2. Catalyst preparation

Catalysts were prepared by wet-impregnation method. First, AMT 1.165 g was added to a 150 mL flask dissolved in 100 mL distilled water until AMT was completely dissolved and 10 g of boehmite was added to the solution and stirred at room temperature for 16 h. The catalyst product was separated by filtration and dried at 383 °C for 12 h. The solid catalyst was crushed into fine powder, following by calcination at 800 °C and 900 °C under a static air atmosphere for 3 h to get WO_x/AlOOH (White Powder).

2%Pt/WO_x/AlOOH was obtained from H_2PtCl_6 0.216 mmol (0.103 g) dissolved in 50 mL distilled water in 100 mL flask and 2 g of boehmite-supported WO_x (WO_x/AlOOH) was added to this solution, following by stirring at room temperature for 16 h, and then evaporation to dryness at 110 °C. The catalyst having pale yellow solid was crushed into fine powder and calcined at 300 °C under a static air atmosphere for 3 h. to obtain 2%Pt/WO_x/AlOOH.

Physical mixing between SiO_2 calcined at 550 °C (1 g) and "AlOOH" calcined at 900 °C (1g) to obtain SiO_2-γ-Al_2O_3 support. Then, AMT (1.165g) solution was impregnated on SiO_2-γ-Al_2O_3 and dried at 110 °C for 12 h, following by calcination at 500 °C for 8 h. After that WO_x/SiO_2-γ-Al_2O_3 was impregnated with H_2PtCl_6 0.216 mmol (0.103 g) to obtain 2%Pt/WO_x/SiO_2-γ-Al_2O_3 catalyst.

2.3. Catalytic reaction procedure

The hydrogenolysis reaction was performed in a 100 mL stainless steel autoclave equipped with Teflon vessel. Typically, 10 mmol of glycerol was added to 30 mL distilled water and 1.5 g of catalyst was placed in vessel. The reactor was stirred at 800 rpm, purged three times with H_2 at 0.5 MPa, and then heated to 140 °C, pressurized to 0.5 MPa for 12 h. After that the reactor was cooled in an ice-water bath and the hydrogen gas was carefully released. The liquid product was analysed by GC

equipped with a flame ionization detector (FID) and a DB-WAX capillary column (30m x 0.32mm x 0.25um).Product analysis
The glycerol conversion and selectivity of product were calculated from the following equation:

$$\text{Conversion (\%)} = \frac{\text{Moles of introduced substrate - Moles of remained substrate}}{\text{Moles of introduced substrate}} \times 100 \quad (1)$$

$$\text{Selectivity (\%)} = \frac{\text{Moles of one product}}{\text{Moles of all product}} \times 100 \quad (2)$$

2.4. Catalyst characterization
The X-ray diffraction (XRD) patterns were analyzed by X- ray diffractometer) Bruker D8 Advance) using Cu Kα irradiation at range between 10 ° and 90 ° with a step of 0.05s-1.The lattice parameter and d-spacing were calculated based on Bragg's law. Crystallite size was calculated by Scherrer equation.
Scanning electron microscope (SEM), the elemental distribution on a surface of the catalysts was investigated with SEM-EDX using Link Isis series 300 program SEM (JEOL model JSM-5800LV).
IR spectra of ammonia adsorption were recorded with Bruker Vertex-70 FT-IR spectrometer equipped with a Harrick Praying Mantis attachment for diffuse reflectance spectroscopy. About 20–25 mg of sample was placed in a Harrick cell. The sample was heated to 500 °C at a heating rate of 10 °C/min under N_2 gas. Holding the temperature for 1 h under H_2/N_2 gas. Then, the sample was heated to 550 °C at a heating rate of 10 °C/min under N_2 gas. The sample was cooled to 40 °C. After that the sample was saturated with 15% NH_3/He for 30 min. The physisorbed ammonia was desorbed in a nitrogen gas flow about 1 h. The spectra were collected using a MCT detector.

3. Results and discussion

3.1. Catalyst characterization
As seen in figure 1, the X-ray diffraction (XRD) patterns did not show any peaks of WO_x species suggesting that tungsten was well dispersed over surface supports. The diffraction peaks of crystalline γ-alumina at 2θ = 37.5°, 45.4, 66.9° [24] were observed. Pt/WO_x/AlOOH catalyst was not reduced in H_2, so the peaks at 2θ values of 46.6°, 54.5° and 80.3° were corresponding to PtO_2 [25].
Figure 2. Shows the SEM image of Pt/WO_x/AlOOH, catalyst that was highly crystalline. Pt and tungsten oxides have good dispersion on surface of support.
IR spectra of ammonia adsorption obtained after NH_3 adsorption are shown in Table 1. According to the literature [23], Bronsted and Lewis acid sites are simply identified and notable by examination of the 1100–1800 cm^{-1} range. The band at 1454 cm^{-1} is characteristic of symmetric deformation mode of the protonated NH_3 (NH^{4+}) coordinated to Brønsted acid. The band at 1260 cm^{-1} is corresponding to the symmetric deformation mode of ammonia coordinated to Lewis acid sites. The total acidity is as follows; Pt/WO_x/γ-Al_2O_3 > Pt/WO_x/SiO_2-γ-Al_2O_3 > Pt/WO_x/AlOOH.
As shown in table 1, the addition of SiO_2 improved Bronsted acidity and the increase in calcination temperature from 800 °C – 900 °C enhanced Bronsted acidity as shown in figure 3.

IOP Conf. Series: Materials Science and Engineering **559** (2019) 012012 doi:10.1088/1757-899X/559/1/012012

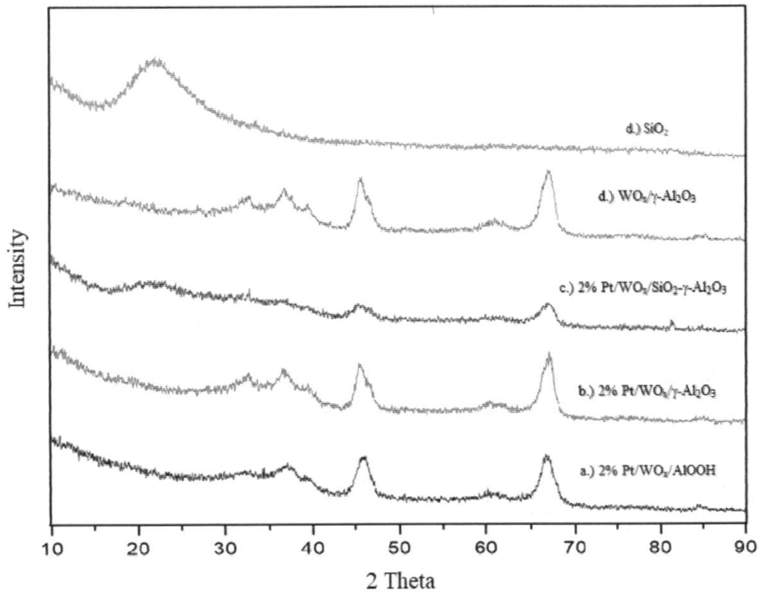

Figure 1. XRD patterns of a) fresh Pt/WO_x/AlOOH, b) Pt/WO_x/AlOOH used, c) WO_x/AlOOH, and d) WO_x/γ-Al_2O_3

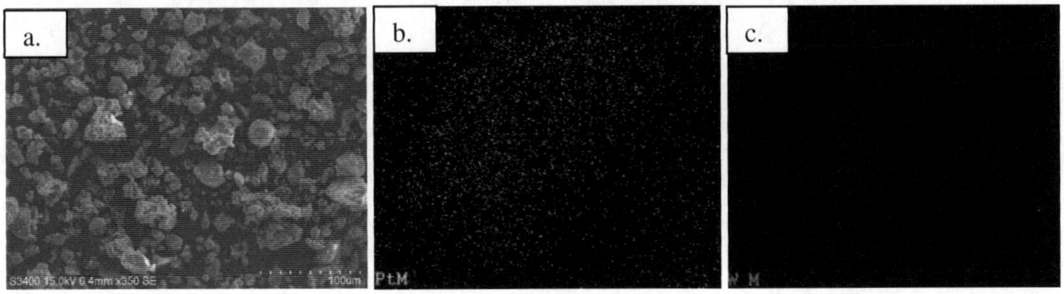

Figure 2. SEM image of a.) Fresh Pt/WO_x/AlOOH catalyst,b.) Pt distribution, and c.) W distribution, obtained from SEM-EDX.

Table 1. The amounts of Bronsted and Lewis acid sites over the catalysts determined from the in situ DRIFTS of adsorbed NH_3

Catalyst	Lewis acid (a.u.)	Bronsted acid (a.u.)	B/L acidity	Total acidity (a.u.)
2%Pt/WO_x/AlOOH	5.3	11.9	2.3	17.2
2%Pt/WO_x/γ-Al_2O_3	5.5	11.2	3.1	22.7
2%Pt/WO_x/SiO_2-γ-Al_2O_3	1.6	19.0	12.1	20.6

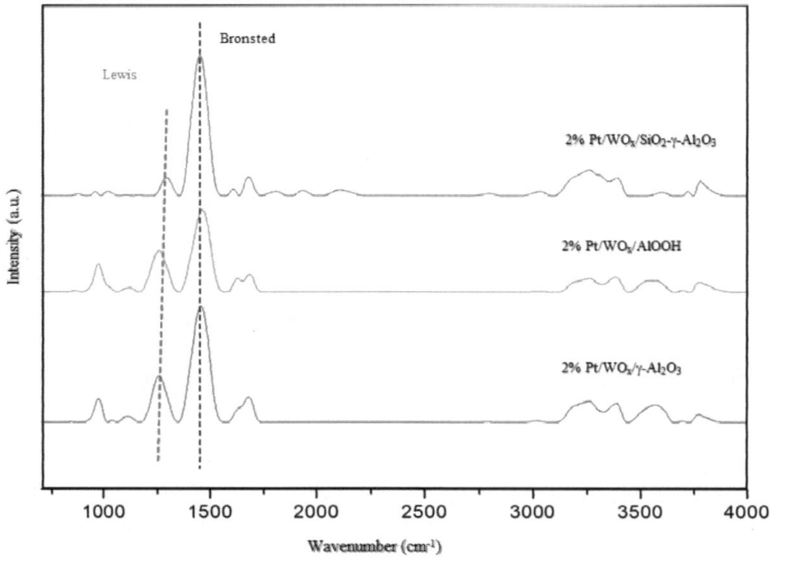

Figure 3. FTIR of adsorbed NH_3 on fresh 2%Pt/WO_x/AlOOH (calcined at 800 °C), fresh 2%Pt/WO_x/γ-Al_2O_3 (calcined at 900 °C) and 2%Pt/WO_x/SiO_2-Al_2O_3 catalysts

3.2. Influence of H_2 pressure

Figure 4. Illustrates the effect of H_2 pressure on hydrogenolysis of glycerol over Pt/WO_x/AlOOH catalyst at 140 °C and 0.5 MPa, which is mildly condition. Glycerol conversion enhanced with increasing of H_2 pressure in range from 0.1-0.5 MPa, associated with the increase of 1,3-PDO selectivity. There is no significant change of 1,2-PDO. The optimal glycerol conversion and selectivity of 1,3-PDO were 32.8% and 18.2% , respectively. In summary, the increase of 1,3-PDO may be due to enhanced hydrogenation rate and raise in the stimulate hydrogen species formed on Pt/WO_x/AlOOH catalyst.[15, 22]

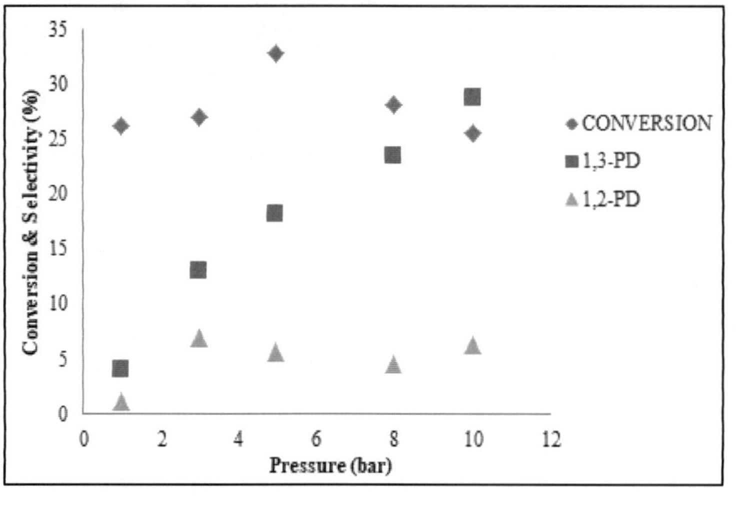

Figure 4. Influence of initial hydrogen pressure on the hydrogenolysis of glycerol over the Pt/WO_x/AlOOH catalyst.

3.3. Influence of support to catalytic activity

Figure 5. Shows the experimental result of hydrogenolysis of glycerol over different catalysts. It was found that Pt/WO_x/SiO_2-γ-Al_2O_3 catalyst gave a remarkable 1,3-PDO selectivity (31.2% at 140° C, 0.5 MPa). Acordding to result of NH_3-IR, Pt/WO_x/SiO_2-γ-Al_2O_3 exhibited the highest Bronsted acid site leading to improved 1,3-PDO selectivity and silica has a strong affinity for glycerol, which was found in the etherification and esterification of glycerol reaction.[15, 20, 26, 27]. The different calcination temperatures affected to increased glycerol conversion and 1,3-PDO selectivity because total acidity and Bronsted acid site increased after calcined at 900 °C (34.6% glycerol conversion, 21.6% 1,3-selectivity). Bronsted acid site plays key role in dehydration of glycerol transfer to 3-HPA as an intermediate chemical in reaction. The hydrogenation of the intermediate on metal sites produces 1,3-PDO.[28]

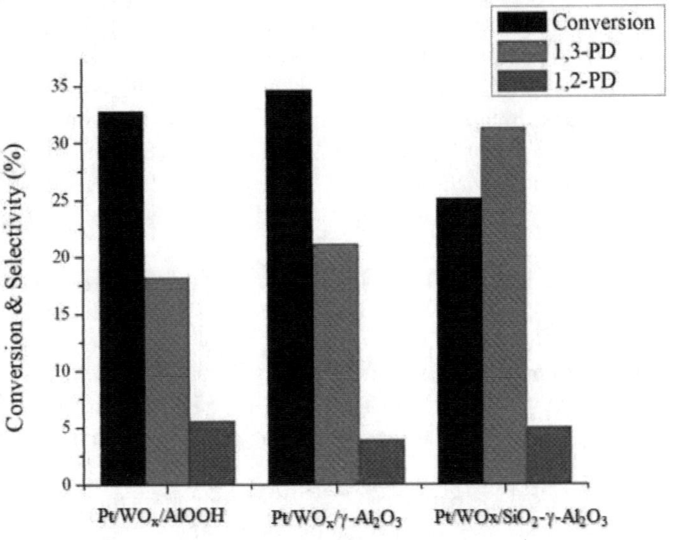

Figure 5. Glycerol conversion and product selectivity obtained from hydrogenolysis of glecerol over diferrent catalysts.

4. Conclusions

Our study has shown that the mild condition to carry out hydrogenolysis of glycerol reaction was selective to form 1,3-PDO depending on Bronsted acid site leading to high 1,3-PDO selectivity. The SiO_2 addition as support notably increased Bronsted acidity and H_2 pressure also influenced on the formation of 1,3-PDO.

5. References

[1] Knothe G, Gerpen, J. V. and Krahl, J.,. The biodiesel handbook 2nd ed Champaing2005.

[2] Singhabhandhu A, Tezuka T. A perspective on incorporation of glycerin purification process in biodiesel plants using waste cooking oil as feedstock. *Energy*. 2010;**35**(6):2493-504.

[3] Wen Z. Application of glycerin. 2012 [Available from: http://www. biodiesel-ua.com/en/bd_glycerin.php.

[4] Bagheri S, Julkapli NM, Yehye WA. Catalytic conversion of biodiesel derived raw glycerol to value added products. *Renewable and Sustainable Energy Reviews*. 2015;**41**:113-27.

[5] Dieuzeide ML, de Urtiaga R, Jobbagy M, Amadeo N. Vapor phase hydrogenolysis of glycerol to 1,2-propanediol at atmospheric pressure over copper catalysts supported on mesoporous alumina. *Catalysis Today*. 2017;**296**:19-25.

[6] Zhou C-H, Beltramini JN, Fan Y-X, Lu GQ. Chemoselective catalytic conversion of glycerol as a biorenewable source to valuable commodity chemicals. *Chemical Society Reviews*. 2008;**37**(3):527-49.

[7] Wang Y, Zhou J, Guo X. Catalytic hydrogenolysis of glycerol to propanediols: a review. *RSC Advances*. 2015;**5**(91):74611-28.

[8] ten Dam J, Hanefeld U. Renewable Chemicals: Dehydroxylation of Glycerol and Polyols. *ChemSusChem*. 2011;**4**(8):1017-34.

[9] Marinoiu A, Ionita G, Gáspár C-L, Cobzaru C, Oprea S. Glycerol hydrogenolysis to propylene glycol. *Reaction Kinetics and Catalysis Letters*. 2009;**97**(2):315-20.

[10] Guo X, Li Y, Song W, Shen W. Glycerol Hydrogenolysis over Co Catalysts Derived from a Layered Double Hydroxide Precursor. *Catalysis Letters*. 2011;**141**(10):1458.

[11] Vasiliadou ES, Eggenhuisen TM, Munnik P, de Jongh PE, de Jong KP, Lemonidou AA. Synthesis and performance of highly dispersed Cu/SiO_2 catalysts for the hydrogenolysis of glycerol. *Applied Catalysis B*: Environmental. 2014;**145**:108-19.

[12] Checa M, Auneau F, Hidalgo-Carrillo J, Marinas A, Marinas JM, Pinel C, et al. Catalytic transformation of glycerol on several metal systems supported on ZnO. *Catalysis Today*. 2012;**196**(1):91-100.

[13] Gallegos-Suarez E, Pérez-Cadenas M, Guerrero-Ruiz A, Rodriguez-Ramos I, Arcoya A. Effect of the functional groups of carbon on the surface and catalytic properties of Ru/C catalysts for hydrogenolysis of glycerol. *Applied Surface Science*. 2013;**287**:108-16.

[14] Kurosaka T, Maruyama H, Naribayashi I, Sasaki Y. Production of 1,3-propanediol by hydrogenolysis of glycerol catalyzed by $Pt/WO_3/ZrO_2$. *Catalysis Communications*. 2008;**9**(6):1360-3.

[15] Zhu S, Zhu Y, Hao S, Chen L, Zhang B, Li Y. Aqueous-Phase Hydrogenolysis of Glycerol to 1,3-propanediol Over $Pt\text{-}H_4SiW_{12}O_{40}/SiO_2$. Catalysis Letters. 2012;**142**(2):267-74.

[16] Nakagawa Y, Shinmi Y, Koso S, Tomishige K. Direct hydrogenolysis of glycerol into 1,3-propanediol over rhenium-modified iridium catalyst. *Journal of Catalysis*. 2010;**272**(2):191-4.

[17] Shinmi Y, Koso S, Kubota T, Nakagawa Y, Tomishige K. Modification of Rh/SiO2 catalyst for the hydrogenolysis of glycerol in water. *Applied Catalysis B*: Environmental. 2010;**94**(3):318-26.

[18] Amada Y, Shinmi Y, Koso S, Kubota T, Nakagawa Y, Tomishige K. Reaction mechanism of the glycerol hydrogenolysis to 1,3-propanediol over Ir–ReOx/SiO2 catalyst. *Applied Catalysis B*: Environmental. 2011;**105**(1):117-27.

[19] Qin L-Z, Song M-J, Chen C-L. Aqueous-phase deoxygenation of glycerol to 1,3-propanediol over Pt/WO3/ZrO2 catalysts in a fixed-bed reactor. Green Chemistry. 2010;12(8):1466-72.

[20] Gong L, Lu Y, Ding Y, Lin R, Li J, Dong W, et al. Selective hydrogenolysis of glycerol to 1,3-propanediol over a $Pt/WO_3/TiO_2/SiO_2$ catalyst in aqueous media. *Applied Catalysis A*: General. 2010;**390**(1):119-26.

[21] Arundhathi R, Mizugaki T, Mitsudome T, Jitsukawa K, Kaneda K. Highly Selective Hydrogenolysis of Glycerol to 1,3-Propanediol over a Boehmite-Supported Platinum/Tungsten Catalyst. *ChemSusChem*. 2013;**6**(8):1345-7.

[22] Nakagawa Y, Tamura M, Tomishige K. Catalytic materials for the hydrogenolysis of glycerol to 1,3-propanediol. *Journal of Materials Chemistry A*. 2014;**2**(19):6688-702.

[23] Wu X, Zhang L, Weng D, Liu S, Si Z, Fan J. Total oxidation of propane on $Pt/WO_x/Al_2O_3$ catalysts by formation of metastable Ptdelta+ species interacted with WO_x clusters. *Journal of hazardous materials*. 2012;**225-226**:146-54.

[24] García-Fernández S, Gandarias I, Requies J, Güemez MB, Bennici S, Auroux A, et al. New approaches to the $Pt/WO_x/Al_2O_3$ catalytic system behavior for the selective glycerol hydrogenolysis to 1,3-propanediol. *Journal of Catalysis*. 2015;**323**:65-75.

[25] Shrotri A, Tanksale A, Beltramini JN, Gurav H, Chilukuri SV. Conversion of cellulose to polyols over promoted nickel catalysts. *Catalysis Science & Technology*. 2012;**2**(9):1852-8.

[26] Gu Y, Azzouzi A, Pouilloux Y, Jérôme F, Barrault J. Heterogeneously catalyzed etherification of glycerol: new pathways for transformation of glycerol to more valuable chemicals. *Green Chemistry*. 2008;**10**(2):164-7.

[27] Karam A, Gu Y, Jérôme F, Douliez J-P, Barrault J. Significant enhancement on selectivity in silica supported sulfonic acids catalyzed reactions. *Chemical Communications*. 2007(22):2222-4.

[28] Zhu S, Qiu Y, Zhu Y, Hao S, Zheng H, Li Y. Hydrogenolysis of glycerol to 1,3-propanediol over bifunctional catalysts containing Pt and heteropolyacids. *Catalysis Today*. 2013;**212**:120-6.

Acknowledgment

The authors appreciatively thanks NRCT-JSPS Joint Research Program (JFY 2018) for the Instrument support and this research is funded by Chulalongkorn University ; Gverment Budget.

Hydrogen Production via Chemical Looping Steam Reforming of Ethanol: Effect of Ce and La on Ni-based Oxygen Carrier Performance

S Isarapakdeetham[1], P Kim-Lohsoontorn[1], S Wongsakulphasatch[2], W Kiatkittipong[3] and S Assabumrungrat [1,a]

[1] Center of Excellence in Catalysis and Catalytic Reaction Engineering, Department of Chemical Engineering, Faculty of Engineering, Chulalongkorn University, Bangkok, 10330, Thailand
[2] Department of Chemical Engineering, Faculty of Engineering, King Mongkut's University of Technology North Bangkok, Bangkok, 10800, Thailand
[3] Department of Chemical Engineering, Faculty of Engineering and Industrial Technology, Silpakorn University, Nakhon Pathom, 73000, Thailand

[a] Corresponding author: suttichai.a@chula.ac.th

Abstract. Chemical looping steam reforming of ethanol for hydrogen production has been studied. In this process, the high hydrogen content compound, ethanol, was reacted with steam and oxygen carriers (OCs) through the redox reaction cycle to produce the clean fuel H_2. The effects of the different promoters, CeO_2 and La_2O_3, in Al_2O_3-supported Ni-based OCs were investigated. The mixed promoters Ni-Al_2O_3 OCs were synthesized by wet impregnation method. The fresh OCs were characterized by X-ray powder diffraction (XRD), N_2 adsorption-desorption, H_2-temperature programmed reduction (H_2-TPR) and Scanning electron microscope/Energy dispersive x-ray spectroscopy (SEM/EDX) technique. The redox activity test was performed in a fixed bed reactor by alternating between fuel and air feeds at isothermal condition of 500°C. The fuel feed was composed of ethanol and steam with steam to ethanol (S/E) mole ratio of 3. It was found that addition of CeO_2 and La_2O_3 on Ni/Al_2O_3 improved the metal dispersion, reduced the crystallite size and enhanced reducibility of NiO. The La_2O_3 promoted (Ni/La_2O_3-Al_2O_3) and CeO_2 and La_2O_3 co-promoted (Ni/La_2O_3-CeO_2-Al_2O_3) OCs showed high performance and stability in chemical looping steam reforming of ethanol with the maximum ethanol conversion of 98.3 and 92.5 %, respectively, with maximum H_2 yield of 2.9 and 3.2 mol/mol ethanol, respectively. In addition, forming of La_2O_3-CeO_2 solid solution phase in Ni/La_2O_3-CeO_2-Al_2O_3 oxygen carriers enhanced the oxygen storage capacity (OSC) and high oxygen mobility (OM) of OCs, resulting in a lower carbon deposition.

1. Introduction

Nowadays, global warming is one of the world's major problems which have drawn great attention from many countries. Greenhouse gases (e.g. CO_2, CH_4, and O_3) generated from human activities especially fossil fuel burning are the main cause of global warming. Since 2006, CO_2 concentration level has been increased from 380 ppm to 408 ppm (https://climate.nasa.gov/). So, all sectors are cooperating to reduce energy consumption and to use more alternative energy such as solar power, wind power, biomass, tidal power, geothermal, etc.

Content from this work may be used under the terms of the Creative Commons Attribution 3.0 licence. Any further distribution of this work must maintain attribution to the author(s) and the title of the work, journal citation and DOI.
Published under licence by IOP Publishing Ltd

Hydrogen has recently been given attention and proposed as one of the most desirable and efficient energy carriers instead of fossil fuels. It is often used as a clean fuel in fuel cells because only water steam is produced with very high energy release, and combustion of hydrogen to produce the energy is not releases the CO_2 [1]. Moreover, hydrogen is used for the conversion of heavy petroleum fractions into lighters by hydrocracking process, for the production of ammonia by Haber process, for the reduction of carbon monoxide to produce methanol. Hydrogen can be produced through different methods. The dominant route to produce hydrogen on a large scale from hydrocarbons or biomass is steam reforming [2-4] while other technologies such as sorption enhanced steam reforming [5, 6] with or without chemical looping process [7-11] from methane [7-9], methanol [12, 13], ethanol [3, 10, 11, 14, 15], glycerol [16, 17], acetic acid [2, 18] have been proposed. Among the hydrocarbons and oxygenated compounds, ethanol is considered as an important candidate because of high hydrogen content, less hazard, safe transport and high heating value, and it can be produced economically and sustainably in large amounts from biomass such as agricultural wastes [19, 20].

There are two groups of metal which are often used as catalysts for steam reforming of ethanol, consisting of noble metals (Rh, Pt, Pd, Ru, Ir) and transition metals (Ni, Co, Cu, Fe). In case of noble metals, there are many advantages for steam reforming comprising of good catalytic activity and resistance to carbon formation except their high cost and low availability [21]. Among the transition metals, Ni is the best catalyst for hydrogen production by steam reforming which shows more than 90% hydrogen selectivity [22]. However, Ni-based catalysts generally suffer rapid deactivation caused by sintering of active nickel species and carbon deposition, which remains a major challenge for developing Ni-based catalysts [23, 24]. An effective way of improving Ni dispersion and suppressing the aggregation of Ni nanoparticles is spreading them on porous supports [25]. Alumina is often used as support because of its high surface area but the Lewis acid sites of the alumina cause carbon deposition due to ethylene production [22, 26]. To adjust the stability of alumina supported Ni, alkali earth metals are promoted on the catalysts to reduce acidity, reduce Ni particle size and improve Ni dispersion [27].

Chemical looping steam reforming refers to the reduction and oxidation of oxygen carriers (OCs) which are circulated between a fuel reactor (FR) and an air reactor (AR). An oxygen carrier, often metal oxide, is used to transfer the oxygen from the air to the fuel between two reactors. The reduced oxygen carrier is then recycled by using air to oxidize the oxygen carrier to its original state. It should be noted that reduced oxygen carrier also acts as a reforming catalyst. The oxidation reaction of OCs is very exothermic and the reduction reaction is endothermic so this process acts as autothermal condition which reduces the external energy consumption [10]. Moreover, carbon deposition on catalyst can be eliminated during the process of OCs regeneration, this process utilizes air rather than pure oxygen from costly air separation process and achieves heat balance in autothermal condition [19].

Metal oxides of Fe, Ni, Co, Cu, Mn, and Cd have been studied, mostly for chemical looping of gaseous fuels such as natural gas or methane [9, 28, 29]. Chemical looping of liquid fuels such as ethanol has also been reported [10, 11, 30]. Metal oxide of Ni is relatively low cost and was found to provide a high oxidation/reduction rate which is suitable for chemical looping process [19, 31]. Zafar *et al.*, (2006) [28] have studied Ni, Cu, Fe and Mn supported on SiO_2 or $MgAl_2O_4$ as OCs and the results showed that Ni-based OCs are the most feasible for chemical looping process. However, Ni is still susceptible to agglomeration and carbon deposition during the reforming process.

For the agglomeration problem, the impregnation of OCs onto substrates was found to reduce agglomeration and increase the reactivity of the OCs [19]. The different types of alumina support were reported to affect the reactivity of the OCs and on the gas product distribution. De Diego *et al.*, (2008) [32] reported that the reduction reactivity of the $NiO/\alpha\text{-}Al_2O_3$ was higher than $NiO/\gamma\text{-}Al_2O_3$ because of the limited solid-state reaction between NiO and $\alpha\text{-}Al_2O_3$. The usage of proper OCs with high oxygen storage capacity (OSC) and high oxygen mobility (OM) was found to reduce carbon deposition on the OCs. Jiang *et al.*, (2018) [33] studied the effect of Ce loading on nickel-phyllosilicate nanotube OCs and found that OCs with high OSC and OM could reduce carbon deposition during hydrogen production process due to the oxygen lattice diffusion to oxidize the carbon formation. Therefore, the excellent OCs should perform as high reactivity, high OSC, high OM, high resistance to sintering and high stability [34].

Rare earth has been widely used as a promoter doped on catalysts in reforming process to improve the metal dispersion and metal-support interaction [26, 35, 36]. Ceria (CeO_{2-x}) is the interesting rare earth metal due to the existing in a fluorite structure, which is favorable for the diffusion of the oxygen lattice and oxygen vacancy formation. Therefore, ceria exhibits high OSC and OM since it easily releases, stores and transports oxygen [37]. Moreover, in reforming process, CeO_2 can promote the water gas shift (WGS) reaction and also prevent the sintering of metallic species. In Ni-ceria/NiO-ceria chemical looping, ceria can also act as an oxygen carrier due to its high OSC and OM properties [38]. La^{3+} aliovalent dopant (La_2O_3) has been used to enhance OSC and OM of ceria. The addition of La_2O_3 is expected to reduce the surface acidity of Al_2O_3 support [39]. Although doped ceria has been employed extensively as a reforming catalyst, up to date the effect of ceria doping as an oxygen carrier on the redox property and stability has not been of much attention in the chemical looping process.

The aim of this study is to study the effect of addition CeO_2 and/or La_2O_3 as a promoter on Ni-based OCs performance for chemical looping steam reforming of ethanol. It is expected that CeO_2 and/or La_2O_3 incorporation ought to improve the Ni dispersion, enhance reducibility of NiO which shortened the dead time in the fuel feed step, improve the efficiency of hydrogen production and reduce carbon deposition during the fuel feed step.

2. Experimental

2.1. Oxygen carrier preparation

The oxygen carriers were prepared by wet impregnation method with an aqueous solution of cerium nitrate hexahydrate ($Ce(NO_3)_3{\cdot}6H_2O$) and Al_2O_3 supports under stirring at the temperature of 70°C until the solution transforms to the slurry. The mixture was then dried at 110°C overnight and calcined at 650°C for 6 hours under static air to obtain CeO_2-Al_2O_3. After that, La_2O_3 was introduced into CeO_2-Al_2O_3 and Al_2O_3 using an aqueous solution of lanthanum nitrate hexahydrate ($La(NO_3)_3{\cdot}6H_2O$) as precursor using a similar procedure as mentioned above obtained La_2O_3-CeO_2-Al_2O_3 and La_2O_3-Al_2O_3.

Supported Ni oxygen carriers were prepared by wet impregnation method using an aqueous solution of nickel nitrate hexahydrate ($Ni(NO_3)_2{\cdot}6H_2O$) stirring with each support (Al_2O_3, CeO_2-Al_2O_3, La_2O_3-Al_2O_3, and La_2O_3-CeO_2-Al_2O_3) using a similar procedure as mentioned above and calcined at 500°C for 4 hours. The Ni, CeO_2 and La_2O_3 content were fixed at 12.5 wt.%, 10 wt.% and 15 wt.%, respectively. These samples were labeled as N/A, N/CA, N/LA and N/LCA where N=Ni, C=CeO_2, L=La_2O_3, and A=Al_2O_3.

2.2. Oxygen carrier characterization

X-ray powder diffraction (XRD) patterns were analyzed, using a Cu Ka radiation as x-ray source at a wavelength of 0.1540 nm (Bruker D8 Advance). The crystallite size of the metal was calculated from the Scherrer's equation. The surface area, pore size and pore volume of the oxygen carrier were calculated according to the Brunauer-Emmet-Teller (BET) method by N_2 adsorption-desorption technique (Micromeritics ASAP 2020 instrument). H_2-TPR experiments (Micromeritics ChemiSorb 2750 instrument) were carried out to determine the strength of the metal-support interaction of the OCs. 0.1 g of each OCs were packed into a quartz tube reactor and heated from room temperature to 800°C at a rate of 10°C/min in 10 % H_2 in Ar. The elemental distribution on a surface and morphology of the catalysts was investigated using Scanning electron microscope/Energy dispersive x-ray spectroscopy (SEM/EDX, Hitachi S-3400N/EDAX Apollo X).

2.3. Activity test

0.5 g OCs supported by quartz wool were loaded into a quartz fixed bed reactor (Ø 16 x 500 mm) to perform chemical looping steam reforming of ethanol. The reactor was equipped with a coaxial thermocouple for temperature monitoring and controlling, and was heated electrically by a tubular furnace to set temperature of 500°C. The schematic diagram of the laboratory system is shown in figure 1.

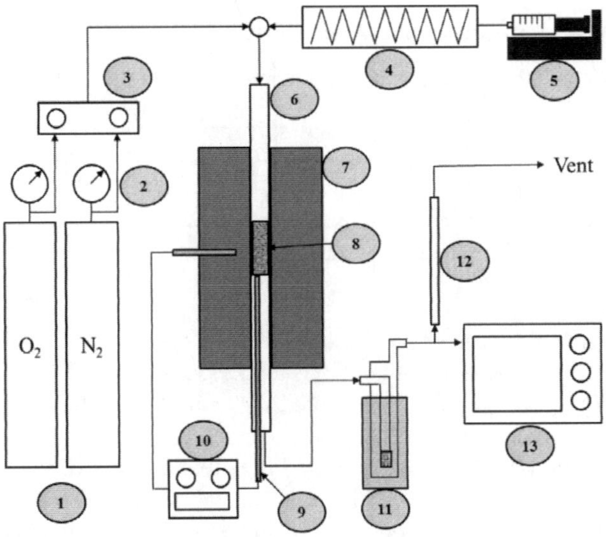

Figure 1. Experimental equipment for hydrogen production.

The test was divided into two steps, fuel (ethanol and steam) feed step and air feed step. In the fuel feed step, the mixture of water and ethanol with steam to ethanol (S/E) mole ratio of 3 was fed from a liquid phase by a syringe pump (KD Scientific Inc.) with a flow rate of 3.4 ml/h and then it was vaporized and mixed with N_2 carrier gas with a total flow rate of 130 ml/min under a WHSV = 6.12 h^{-1}. The outlet stream was trapped by an ice bath and was analyzed by gas chromatography (Shimadzu Inc.). After the fuel feed step, the gas products were removed by N_2 purge until no residue gas was measured. The air feed step was then performed by flowing air at a flow rate of 100 ml/min to oxidize metal into metal oxide. The exit gas in the air feed step was collected and analyzed by gas chromatography. The activity of OCs was calculated in terms of ethanol conversion ($X_{ethanol}$), hydrogen yield (Y_{H2}) and selectivity of products (Sel. i) as follow:

$$X_{ethanol} = \frac{\dot{n}_{ethanol,in} - \dot{n}_{ethanol,out}}{\dot{n}_{ethanol,in}} \times 100 \quad (1)$$

$$Y_{H_2} = \frac{\dot{n}_{H_2,out}}{\dot{n}_{ethanol,in}} \quad (2)$$

$$Sel.\,i = \frac{\dot{n}_{i,out}}{\sum \dot{n}_{i,out}} \quad (3)$$

where ṅ is the relevant molar flows (mol/min), i represents CO, CO_2, CH_4, and CH_3CHO

3. Results and discussion

3.1. Characterization of oxygen carrier

Table 1 shows surface area, pore volume, pore size and NiO crystallite size of four OCs. When compared with Al_2O_3, Ni/Al_2O_3 showed higher surface area, larger pore volume and smaller pore size due to the pore blocking occurring from a large NiO crystallite size. For the modified OCs, CeO_2 and/or La_2O_3 was found to improve the physical properties of Ni/Al_2O_3, leading to higher surface area, pore volume and pore size. The crystallite size of NiO are shown in the following order: Ni/Al_2O_3 > Ni/CeO_2-Al_2O_3 > Ni/La_2O_3-CeO_2-Al_2O_3 > Ni/La_2O_3-Al_2O_3. The pore size of Ni/CeO_2-

Al_2O_3, Ni/La_2O_3-Al_2O_3, and Ni/La_2O_3-CeO_2-Al_2O_3 were larger than Ni/Al_2O_3 due to a smaller NiO crystallite size, resulting in a less pore blocking.

Table 1. Physical properties of oxygen carriers.

Oxygen carriers (OCs)	Surface area (m^2/g)	Pore volume (cm^3/g)	Pore size (nm)	NiO crystallite size (nm)
A	1.618	0.000325	37.33	-
N/A	3.078	0.006961	9.44	39.40
N/CA	6.407	0.023791	14.20	34.35
N/LA	6.069	0.025811	21.90	21.77
N/LCA	7.315	0.033301	21.20	27.97

The H_2-TPR technique was used to identify the type of reducible species formed in the OCs after the calcination process. The three main reduction peaks obtained from H_2-TPR profiles (Figure 2) in the range of 230 – 350, 350 – 450 and 450 – 550°C which represented the reduction of bulk NiO; NiO species having weak to medium interaction with support; and, NiO species having strong interaction with support, respectively [26]. The modified Ni/Al_2O_3 showed the lower temperature of the first and second reduction peak, indicating more Ni species content with weaker metal-support interaction (MSI) [19]. The third peak of Ni/La_2O_3-Al_2O_3 and Ni/La_2O_3-CeO_2-Al_2O_3 shifted to the higher temperature compared to Ni/Al_2O_3, indicating the stronger MSI due to relatively smaller NiO crystallite size and larger specific surface area, as shown in table 1. The addition of CeO_2 and/or La_2O_3 on Ni/Al_2O_3 reduced the NiO crystallite size dispersed on the surface and reduced MSI obtained more bulk NiO which was easier to reduce NiO species to formed Ni species in the fuel feed step.

XRD patterns of the OCs after the calcination process are shown in figure 3. For all OCs, the reflections of NiO and Al_2O_3 are present at 2Θ = 37.3, 43.3, 62.9, 75.4, 79.32 (JCPDS 73-1523) and 25.6, 35.2, 37.8, 52.5, 57.5, 61.3, 66.5, 68.2, 76.9, respectively [19]. The peaks observed at 2Θ of 28.6, 33.2, 47.6 and 56.4 were indexed as (111), (200), (220) and (311) planed of fluorite structure of CeO_2 (JCPDS 81-0792) [36]. The small peaks of La_2O_3 were located on 26.8, 40.6 and 53.7, indicating the high dispersion on the surface of Al_2O_3 [40, 41].

For the La_2O_3-CeO_2 doping (figure 2(d)), the peaks of CeO_2 on the location of 2Θ = 28.6, 33.2, 47.6 and 56.4 were shifted to lower angle due to the formation of La_2O_3-CeO_2 solid-solution, corresponding to the work of Pramila et al. (2016) [42] which reported that the partial substitution of La^{3+} (0.11nm) and Ce^{4+} (0.09 nm) leads to solid solution formation and the lattice shape distortion. In order to maintain electrical neutrality in the La_2O_3-CeO_2 solid-solution crystal structure, oxygen ion vacancies will be created, in turn increasing OSC and OM [43]. The SEM micrographs and EDX mapping of Ni of fresh OCs are shown in figure 4. The morphology of fresh OCs showed the uniform dispersion of metal on the surface. Particle size from SEM micrographs could be related to the crystallite size reported in table 1.

3.2. Activity test

The activity test of OCs was evaluated by the ethanol conversion and hydrogen yield (figure 5-6). For ethanol conversion, all OCs presented the same trend. Ethanol conversion and hydrogen yield were initially increased since NiO was reduced by ethanol to formed metallic Ni which acts as catalyst for steam reforming. After that ethanol conversion and hydrogen yield decreased, likely due to deactivation of metallic Ni through carbon deposition. The overall ethanol conversion was shown in the following order of Ni/La_2O_3-Al_2O_3 = Ni/La_2O_3-CeO_2-Al_2O_3 > Ni/CeO_2-Al_2O_3 > Ni/Al_2O_3. The Ni/La_2O_3-Al_2O_3 and Ni/La_2O_3-CeO_2-Al_2O_3 exhibited relatively smaller NiO crystallite size and larger surface area which facilitated the breakage of C-C and C-H bonds [19].

IOP Conf. Series: Materials Science and Engineering **559** (2019) 012013 doi:10.1088/1757-899X/559/1/012013

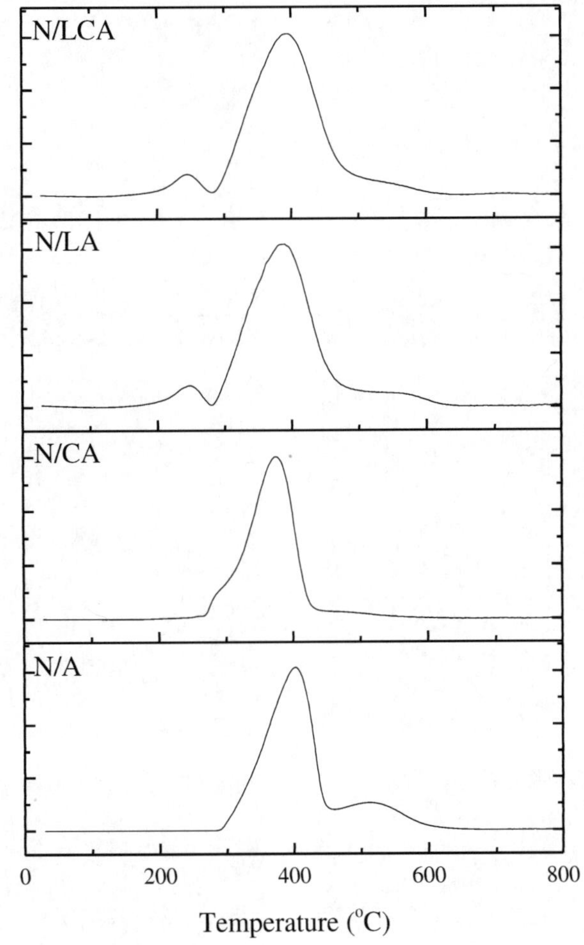

Figure 2. H_2-TPR profiles of the OCs.

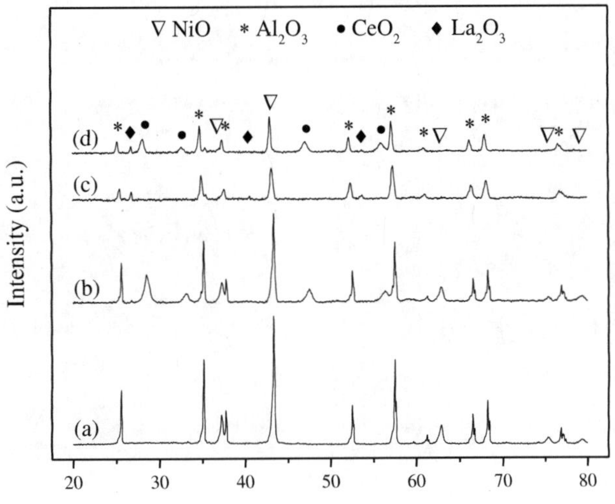

Figure 3. XRD patterns of (a) N/A (b) N/CA (c) N/LA and (d) N/LCA OCs.

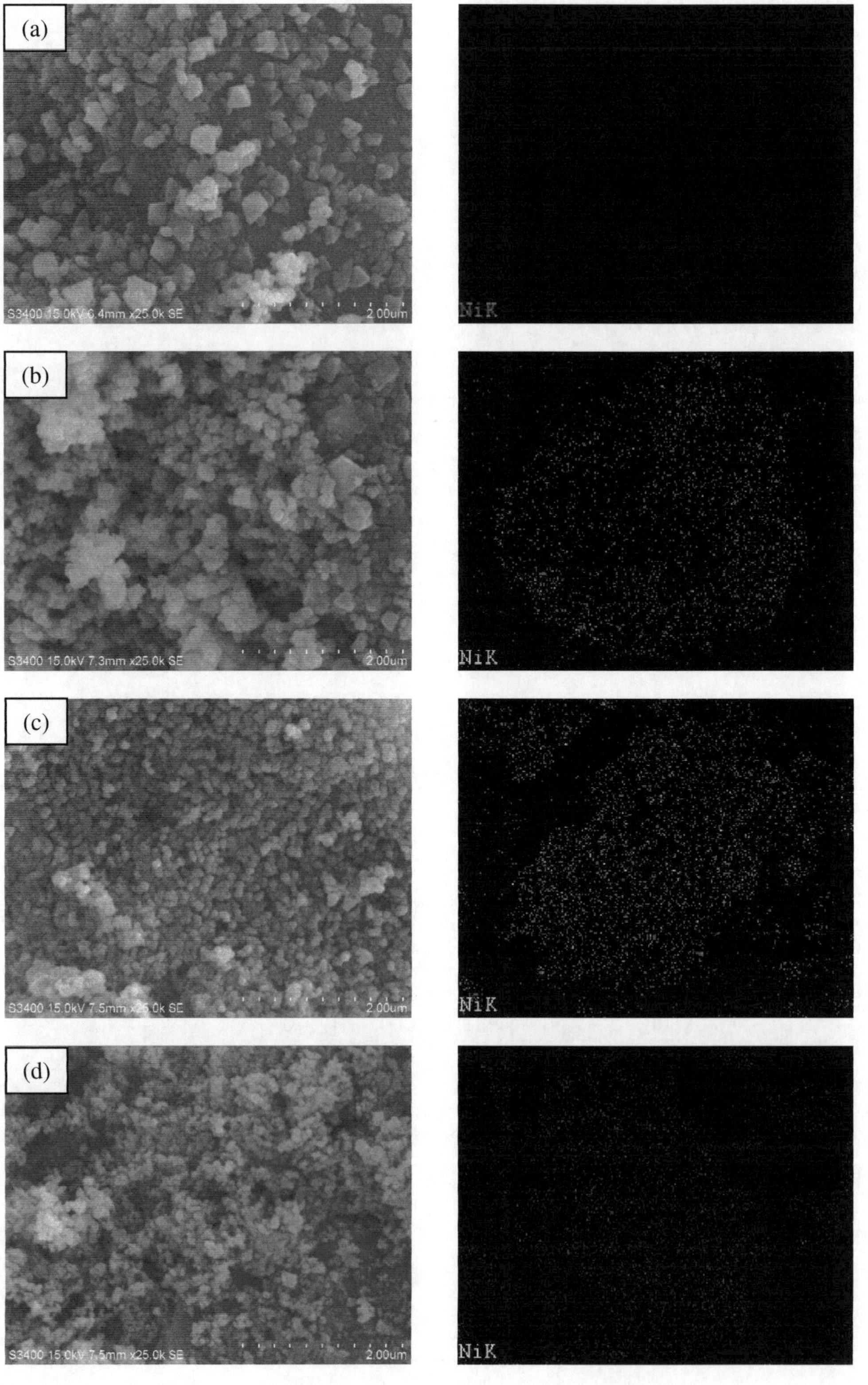

Figure 4. SEM profiles (left) and EDX mapping of Ni (right) of (a) N/A (b) N/CA (c) N/LA and (d) N/LCA OCs.

During the initial fuel feed step, there was no H_2 generation which was called 'dead time'. Unless NiO was sufficiently reduced to Ni, steam reforming of ethanol would not occur. The dead time was considered as a significant indicator for the redox performance of OCs. Figure 6(a) showed H_2 yield of the initial 5 minutes of the fuel feed step. The results showed that Ni/La_2O_3-Al_2O_3 and Ni/La_2O_3-CeO_2-Al_2O_3 were easily reduced, corresponding to the H_2-TPR results (Figure 2). The overall H_2 yield (Figure 6(b)) within 3 hours showed in the following order of Ni/La_2O_3-CeO_2-Al_2O_3 > Ni/La_2O_3-Al_2O_3 > Ni/CeO_2-Al_2O_3 > Ni/Al_2O_3. For the air feed step, air was fed into the reactor, oxygen in air would oxidize carbon deposition on OCs to form CO_2 as shown in figure 7. In the initial of air feed step, the results showed that there was the same level of CO_2 in the gas outlet before the amount of CO_2 from Ni/La_2O_3-Al_2O_3 and Ni/La_2O_3-CeO_2-Al_2O_3 OCs decreased within around 40 minutes. On the other hand, Ni/Al_2O_3 OCs showed a higher amount of CO_2 and remained for 50 minutes. Ni/La_2O_3-Al_2O_3 and Ni/La_2O_3-CeO_2-Al_2O_3 showed relatively lowest carbon deposition and faster regeneration process. Addition of CeO_2 and/or La_2O_3 on Ni/Al_2O_3 helped decrease carbon deposition and also decrease OCs regeneration duration. This was likely due to an increased OSC and OM which facilitated oxygen lattice diffuse through the structure of OCs and oxidized carbon formed on OCs during the fuel feed step.

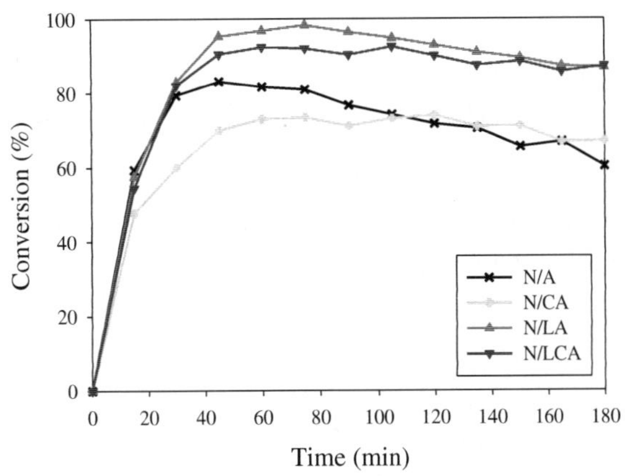

Figure 5. Ethanol conversion during fuel feed step.

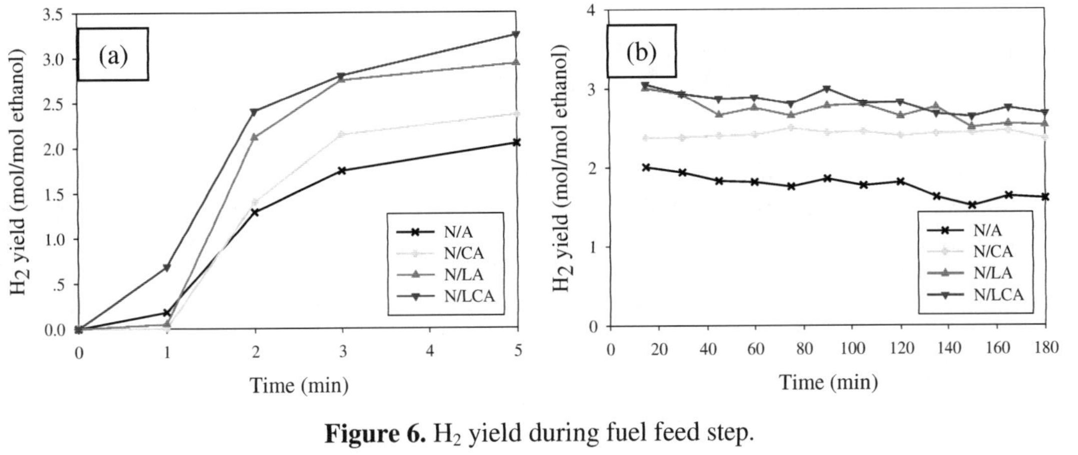

Figure 6. H_2 yield during fuel feed step.

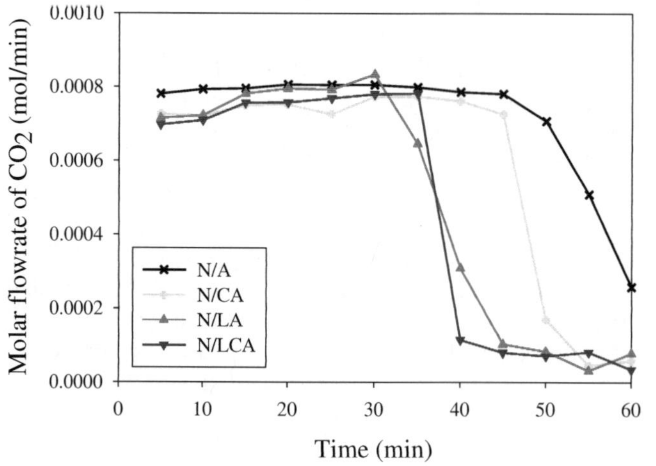

Figure 7. Molar flow rate of CO_2 during air feed step.

4. Conclusion

The addition of rare earth oxide (CeO_2, La_2O_3) on 12.5 wt.% Ni/Al_2O_3 improved the metal dispersion with a small crystallite size of NiO and enhanced reducibility of bulk NiO. The smallest NiO crystallite size and large surface area were obtained by Ni/La_2O_3-Al_2O_3 and Ni/La_2O_3-CeO_2-Al_2O_3, respectively. The La_2O_3-CeO_2 solid solution was formed in Ni/La_2O_3-CeO_2-Al_2O_3 OCs enhancing the OSC and OM. The high OSC and OM influenced to lower carbon deposition on OCs. The best performance of OCs for chemical looping steam reforming of ethanol were Ni/La_2O_3-Al_2O_3 and Ni/La_2O_3-CeO_2-Al_2O_3 which obtained high ethanol conversion, high H_2 yield and low carbon deposition.

5. References

[1] Nguyen VN, Deja R, Peters R, Blum L, Stolten D. Study of the catalytic combustion of lean hydrogen-air mixtures in a monolith reactor. *International Journal of Hydrogen Energy*. 2018;**43**(36):17520-30.

[2] Zhang C, Hu X, Zhang Z, Zhang L, Dong D, Gao G, et al. Steam reforming of acetic acid over Ni/Al2O3 catalyst: Correlation of calcination temperature with the interaction of nickel and alumina. *Fuel*. 2018;**227**:307-24.

[3] Dobosz J, Małecka M, Zawadzki M. Hydrogen generation via ethanol steam reforming over Co/HAp catalysts. *Journal of the Energy Institute*. 2018;**91**(3):411-23.

[4] Parlett CMA, Aydin A, Durndell LJ, Frattini L, Isaacs MA, Lee AF, et al. Tailored mesoporous silica supports for Ni catalysed hydrogen production from ethanol steam reforming. *Catalysis Communications*. 2017;**91**:76-9.

[5] Shokrollahi Yancheshmeh M, Radfarnia HR, Iliuta MC. High temperature CO2 sorbents and their application for hydrogen production by sorption enhanced steam reforming process. *Chemical Engineering Journal*. 2016;**283**:420-44.

[6] Xie H, Yu Q, Zuo Z, Han Z, Yao X, Qin Q. Hydrogen production via sorption-enhanced catalytic steam reforming of bio-oil. *International Journal of Hydrogen Energy*. 2016;**41**(4):2345-53.

[7] Shen Y, Zhao K, He F, Li H. The structure-reactivity relationships of using three-dimensionally ordered macroporous LaFe1−xNixO3 perovskites for chemical-looping steam methane reforming. *Journal of the Energy Institute*. 2018.

[8] Ding H, Xu Y, Luo C, Wang Q, Shen C, Xu J, et al. A novel composite perovskite-based material for chemical-looping steam methane reforming to hydrogen and syngas. *Energy Conversion and Management*. 2018;**171**:12-9.

[9] Meshksar M, Daneshmand-Jahromi S, Rahimpour MR. Synthesis and characterization of cerium promoted Ni/SBA-16 oxygen carrier in cyclic chemical looping steam methane reforming. *Journal of the Taiwan Institute of Chemical Engineers*. 2017;**76**:73-82.

[10] Wang W, Fan L, Wang G. Study on chemical looping reforming of ethanol (CLRE) for hydrogen production using NiMn 2 O 4 spinel as oxygen carrier. *Journal of the Energy Institute*. 2017;**90**(6):884-92.

[11] Zhang Q, Li L, Jiang B, Tang D, Dou B. Hydrogen by chemical looping reforming of ethanol: The effect of promoters on La2-xMxNiO4-λ (M = Ca, Sr and Ce) oxygen carriers. *Chemical Engineering Science*. 2017;**174**:259-67.

[12] Lei Y, Luo Y, Li X, Lu J, Mei Z, Peng W, et al. The role of samarium on Cu/Al2O3 catalyst in the methanol steam reforming for hydrogen production. *Catalysis Today*. 2018;**307**:162-8.

[13] González-Gil R, Herrera C, Larrubia MÁ, Kowalik P, Pieta IS, Alemany LJ. Hydrogen production by steam reforming of DME over Ni-based catalysts modified with vanadium. *International Journal of Hydrogen Energy*. 2016;**41**(43):19781-8.

[14] Santander JA, Tonetto GM, Pedernera MN, López E. Ni/CeO2–MgO catalysts supported on stainless steel plates for ethanol steam reforming. *International Journal of Hydrogen Energy*. 2017;**42**(15):9482-92.

[15] Li L, Song Y, Jiang B, Wang K, Zhang Q. A novel oxygen carrier for chemical looping reforming: LaNiO3 perovskite supported on montmorillonite. *Energy*. 2017;**131**:58-66.

[16] Aman D, Radwan D, Ebaid M, Mikhail S, van Steen E. Comparing nickel and cobalt perovskites for steam reforming of glycerol. *Molecular Catalysis*. 2018;**452**:60-7.

[17] Bepari S, Pradhan NC, Dalai AK. Selective production of hydrogen by steam reforming of glycerol over Ni/Fly ash catalyst. *Catalysis Today*. 2017;**291**:36-46.

[18] Li L, Jiang B, Tang D, Zhang Q, Zheng Z. Hydrogen generation by acetic acid steam reforming over Ni-based catalysts derived from La1−xCexNiO3 perovskite. *International Journal of Hydrogen Energy*. 2018;**43**(14):6795-803.

[19] Wang K, Dou B, Jiang B, Zhang Q, Li M, Chen H, et al. Effect of support on hydrogen production from chemical looping steam reforming of ethanol over Ni-based oxygen carriers. *International Journal of Hydrogen Energy*. 2016;**41**(39):17334-47.

[20] Wang K, Dou B, Jiang B, Song Y, Zhang C, Zhang Q, et al. Renewable hydrogen production from chemical looping steam reforming of ethanol using xCeNi/SBA-15 oxygen carriers in a fixed-bed reactor. *International Journal of Hydrogen Energy*. 2016;**41**(30):12899-909.

[21] Sharma YC, Kumar A, Prasad R, Upadhyay SN. Ethanol steam reforming for hydrogen production: Latest and effective catalyst modification strategies to minimize carbonaceous deactivation. *Renewable and Sustainable Energy Reviews*. 2017;**74**:89-103.

[22] Contreras JL, Salmones J, Colín-Luna JA, Nuño L, Quintana B, Córdova I, et al. Catalysts for H2 production using the ethanol steam reforming (a review). *International Journal of Hydrogen Energy*. 2014;**39**(33):18835-53.

[23] Zhang C, Li S, Wu G, Gong J. Synthesis of stable Ni-CeO2 catalysts via ball-milling for ethanol steam reforming. *Catalysis Today*. 2014;**233**:53-60.

[24] Sehested J. Four challenges for nickel steam-reforming catalysts. *Catalysis Today*. 2006;**111**(1-2):103-10.

[25] Cao A, Lu R, Veser G. Stabilizing metal nanoparticles for heterogeneous catalysis. *Phys Chem Chem Phys*. 2010;**12**(41):13499-510.

[26] Osorio-Vargas P, Flores-González NA, Navarro RM, Fierro JLG, Campos CH, Reyes P. Improved stability of Ni/Al2O3 catalysts by effect of promoters (La2O3, CeO2) for ethanol steam-reforming reaction. *Catalysis Today*. 2016;**259**:27-38.

[27] Song JH, Han SJ, Yoo J, Park S, Kim DH, Song IK. Hydrogen production by steam reforming of ethanol over Ni–X/Al2O3–ZrO2 (X = Mg, Ca, Sr, and Ba) xerogel catalysts: Effect of alkaline earth metal addition. *Journal of Molecular Catalysis A: Chemical*. 2016;**415**:151-9.

[28] Zafar Q, Mattisson T, Gevert B. Redox Investigation of Some Oxides of Transition-State Metals Ni, Cu, Fe, and Mn Supported on SiO2 and MgAl2O4. *Energy & Fuels*. 2006; **20**:*34-44.*

[29] Karimi E, Forutan HR, Saidi M, Rahimpour MR, Shariati A. Experimental Study of Chemical-Looping Reforming in a Fixed-Bed Reactor: Performance Investigation of Different Oxygen Carriers on Al2O3 and TiO2 Support. *Energy & Fuels*. 2014;**28**(4):2811-20.
[30] Trevisanut C, Vozniuk O, Mari M, Urrea SYA, Lorentz C, Millet J-MM, et al. The Chemical-Loop Reforming of Alcohols on Spinel-Type Mixed Oxides: Comparing Ni, Co, and Fe Ferrite vs Magnetite Performances. *Topics in Catalysis*. 2016;**59**(17-18):1600-13.
[31] Rubel A, Liu K, Neathery J, Taulbee D. Oxygen carriers for chemical looping combustion of solid fuels. *Fuel*. 2009;**88**(5):876-84.
[32] de Diego LF, Ortiz M, Adánez J, García-Labiano F, Abad A, Gayán P. Synthesis gas generation by chemical-looping reforming in a batch fluidized bed reactor using Ni-based oxygen carriers. *Chemical Engineering Journal*. 2008;**144**(2):289-98.
[33] Jiang B, Li L, Bian Z, Li Z, Othman M, Sun Z, et al. Hydrogen generation from chemical looping reforming of glycerol by Ce-doped nickel phyllosilicate nanotube oxygen carriers. *Fuel*. 2018;**222**:185-92.
[34] Adanez J, Abad A, Garcia-Labiano F, Gayan P, de Diego LF. Progress in Chemical-Looping Combustion and Reforming technologies. *Progress in Energy and Combustion Science*. 2012;**38**(2):215-82.
[35] Li D, Zeng L, Li X, Wang X, Ma H, Assabumrungrat S, et al. Ceria-promoted Ni/SBA-15 catalysts for ethanol steam reforming with enhanced activity and resistance to deactivation. *Applied Catalysis B: Environmental*. 2015;**176-177**:532-41.
[36] Dan M, Mihet M, Tasnadi-Asztalos Z, Imre-Lucaci A, Katona G, Lazar MD. Hydrogen production by ethanol steam reforming on nickel catalysts: Effect of support modification by CeO2 and La2O3. *Fuel*. 2015;**147**:260-8.
[37] Cheng Z, Qin L, Fan JA, Fan L-S. New Insight into the Development of Oxygen Carrier Materials for Chemical Looping Systems. *Engineering*. 2018;**4**(3):343-51.
[38] Löfberg A, Guerrero-Caballero J, Kane T, Rubbens A, Jalowiecki-Duhamel L. Ni/CeO2 based catalysts as oxygen vectors for the chemical looping dry reforming of methane for syngas production. *Applied Catalysis B: Environmental*. 2017;**212**:159-74.
[39] Charisiou ND, Tzounis L, Sebastian V, Hinder SJ, Baker MA, Polychronopoulou K, et al. Investigating the correlation between deactivation and the carbon deposited on the surface of Ni/Al2O3 and Ni/La2O3-Al2O3 catalysts during the biogas reforming reaction. *Applied Surface Science*. 2018.
[40] Sanchezsanchez M, Navarro R, Fierro J. Ethanol steam reforming over Ni/La–Al2O3 catalysts: Influence of lanthanum loading. *Catalysis Today*. 2007;**129**(3-4):336-45.
[41] Xu J, Zhou W, Wang J, Li Z, Ma J. Characterization and Analysis of Carbon Deposited during the Dry Reforming of Methane over Ni/La2O3/Al2O3 Catalysts. *Chinese Journal of Catalysis*. 2009;**30**(11):1076-84.
[42] Patil P, Selvakumar D, Kumar NS. Size Controlled Preparation and Effect of Sintering Temperature on the Structural Properties of Nanocrystalline Ceria-Lanthana Solid Solution. *Materials Today: Proceedings*. 2016;**3**(6):1712-7.
[43] Li F, Sun Z, Luo S, Fan L-S. Ionic diffusion in the oxidation of iron—effect of support and its implications to chemical looping applications. *Energy & Environmental Science*. 2011;**4**(3):876.

Acknowledgements
The authors would like to acknowledge the joint project "The National Research Council of Thailand (NRCT) and The National Natural Science Foundation of China (NSFC)" for funding support.

IOP Conf. Series: Materials Science and Engineering **559** (2019) 012014 doi:10.1088/1757-899X/559/1/012014

Effect of Water Content in Waste Cooking Oil on Biodiesel Production via Ester-transesterification in a Single Reactive Distillation

Ajala Anantapinitwatna[1], Kanokwan Ngaosuwan[2,a], Worapon Kiatkittipong[3], Doonyapong Wongsawaeng[4], Suttichai Assabumrungrat[1]

[1]Center of Excellence on Catalysis and Catalytic Reaction Engineering, Department of Chemical Engineering, Faculty of Engineering, Chulalongkorn University, Bangkok, 10330, Thailand

[2]Division of Chemical Engineering, Faculty of Engineering, Rajamangala University of Technology Krungthep, Bangkok, 10120, Thailand

[3]Department of Chemical Engineering, Faculty of Engineering and Industrial Technology, Silpakorn University, Nakhon Pathom 73000, Thailand

[4]Department of Nuclear Engineering, Faculty of Engineering, Chulalongkorn University, Bangkok, 10330, Thailand

Corresponding author: kanokwan.n@rmutk.ac.

Abstract. A single-column reactive distillation is employed to convert waste cooking oil to biodiesel via simultaneous esterification and transesterification reactions. The waste cooking oil contained 10 wt% of free fatty acid (FFA) and the contaminated water (2-8 wt%) was used as feedstocks for biodiesel production. Esterification was occurred in the upper of column to reduce FFA content to less than 1 wt% and to prevent the simultaneous saponification by base catalyst. Moreover, the contaminated water and by-product water (from esterification) together was removed in-situ as the distillate product. The reactive distillation has been optimally designed to have 16 stages consisting of 5 esterification, 9 transesterification stages and each stage for reboiler and condenser. The feed locations of oil and methanol were at the top of the column and at stage 6, respectively. The optimum methanol to oil ratio was 6:1 with reflux ratio 0.1 while the reboiler duty was in the range of 100-150 kW depended on the water content. This condition successfully converts 97 wt% of waste cooking oil to biodiesel with 96.5% purity.

1. Introduction

Nowadays, the energy requirement is continuously increasing; while, fossil fuel is decreasing and global warming is becoming an important issue. Therefore, the development of the clean and renewable energy to replace petroleum-based fuel is necessary.

Biodiesel is one of the alternatives to replace petroleum-based diesel fuel because the similarity of their properties. The major feedstock used in the production of biodiesel are vegetable oils, animal fats and waste cooking oils (WCO). Typically, more than 70% of the production costs of biodiesel are dependent on the price feedstocks [1]. Low-cost raw material such as waste cooking oil (WCO) can reduce around 50% of the biodiesel cost comparing to the use of fresh palm oil, especially in Thailand. However, major drawback of using WCO as a feedstock is due to its typically high free fatty acid

Content from this work may be used under the terms of the Creative Commons Attribution 3.0 licence. Any further distribution of this work must maintain attribution to the author(s) and the title of the work, journal citation and DOI.
Published under licence by IOP Publishing Ltd

(FFA) and water content. FFA content in WCO can react with base catalyst via saponification as displayed in figure 1, which consumes the catalyst resulting in lower yield of biodiesel.

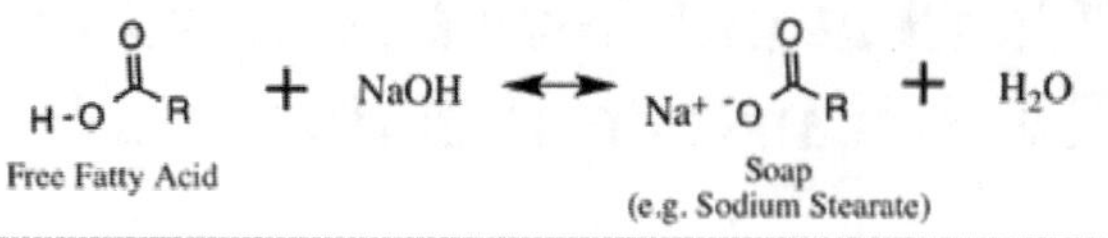

Figure 1. Saponification of FFA.

To address this problem, several researchers have proposed a two-step process in which the WCO is first undergone esterification using acid catalyst (Figure 2) to lower the FFA content below 1wt% following by base-catalysed transesterification [1-3]. In this work, the heterogeneous catalyst, Amberlyst-15 resin and CaO/Al_2O_3 are applied.

$$\underset{\text{Free fatty acid}}{\text{R-COOH}} + \underset{\text{Alcohol}}{\text{R'-OH}} \rightleftharpoons \underset{\text{Ester (biodiesel)}}{\text{R-COO-R'}} + \underset{\text{Water}}{H_2O}$$

Figure 2. Esterification of FFA. [4]

The heterogeneous catalysis can overcome homogeneous catalysis by elimination of the washing section and the reduction of huge amounts of waste water [5]. Biodiesel production process can be separated into two main parts: the reaction and the separation parts. The combination of reaction and separation within single unit operation is called reactive distillation (RD). This unit is specially designed for the equilibrium limited reactions and consecutive reactions. The RD has been utilized for reactions such as esterification, ester-hydrolysis reaction, transesterification, and etherification [6]. Recently, Petchsoongsakul et al. [7] developed the hybridization of ester-transesterification for biodiesel production in a single RD column. This process used heterogeneous catalyst including of Amberlyst-15 and CaO/Al_2O_3. The pseudo-homogeneous kinetic model was used to simulate the combined ester-transesterifications which could not take into account the effect of water. When the esterification takes place, a by-product water is also produced.

Moreover, the raw WCO has around 2-5 wt% water contaminated. It was found that the presence of water could influence on the catalyst activity. [6, 8, 9]. Therefore, the water content should be accounted into kinetic rate model. Therefore, the kinetic model needs to consider the water content to investigate the effect of water on the performance of the single hybridized reactive distillation. Steinigeweg and Gmehling [6] reported that the swelling and poisoning of Amberlyst-15 (a strong acidic ion-exchange resin catalyst) were observed in the presence of water and proposed the kinetic rate model of esterification of the fatty acid decanoic acid and methanol using Amberlyst-15 as a catalyst. The sorption of water was accounted into this kinetic model using a Langmuir-Hinshelwood-Hougen-Watson (LHHW) approach. The kinetic model proposed by Steinigeweg and Gmehling [6] has been used to simulate for transesterification at the bottom of column.

Therefore, this work aims to simulate the effect of water on the catalytic activity via esterification and transesterification using WCO on the number of stages, reflux ratio, methanol to oil molar ratio and feed location of the single reactive distillation. The mixture of oleic acid and triolein was used as a model compound for WCO as presented in table 1. The proposed kinetic rate model of esterification of the fatty acid as decanoic acid and methanol [6] was verified and validated prior to employ for esterification of oleic acid and methanol.

IOP Conf. Series: Materials Science and Engineering **559** (2019) 012014 doi:10.1088/1757-899X/559/1/012014

Table 1. Composition of WCO.

	Model compound	wt%
Triglyceride	Triolein	82-88
FFA	Oleic acid	10
Water	DI-water	2-8

2. Kinetic experimental

2.1. Kinetic Modeling

Steinigeweg and Gmehling [6] proposed the LHHW equation for a reversible esterification using an ion-exchange resin (Amberlyst-15) as a catalyst (eq. (1));

$$r = \frac{1}{v_i}\frac{dn_i}{dt} = m_{cat}\left(\frac{k_1^* K_o leica_o leicK_M eOHa_M eOH - k_{(-1)}^* K_M eOlea_M eOleK_{H_2O}a_{H_2O}}{1+(K_{Oleic}a_{Oleic} + K_{MeOH}a_{MeOH} + K_{MeOle}a_{MeOle} + K_{H_2O}a_{H_2O})^2}\right) \quad (1)$$

where the activities were used instead of concentrations or mole fraction. Water is preferably adsorbate and therefore the sorption effects can be summarized with a singular sorption constant K_{Sorb}. Then the eq. (1) can be simplified as following eq. (2).

$$r = \frac{1}{v_i}\frac{dn_i}{dt} = m_{cat} \times \left(\frac{k_1 a_{Oleic}a_{MeOH}}{(K_{Sorb}a_{H_2O})^2} - \frac{k_{(-1)}a_{MeOle}}{K_{Sorb}a_{H_2O}}\right) \quad (2)$$

2.2. Materials ormatting

Oleic acid was obtained from Vicchi enterprise in Thailand. Analytical grade methanol (99.8% purity) from Ajax FineChem is used as reactant of esterification reaction. Amberlyst-15 ion-exchange resin from Sigma-Aldrich was dried at 110 °C overnight then kept in desiccator before used. Methyl valerate and 2-propanol used as internal standard and solvent, respectively for GC analysis were provided by Sigma-Aldrich

2.3. Esterification experiment

The experiments were performed in a 3-neck-flask with condenser and thermometer into an oil bath to control temperature of reaction mixture at 60 °C using a stirring speed of 600 rpm. Prior to the experiment, both reactants were pre-heated to the desired temperature of 60 °C in a separate tank. Then, the reaction time was started when the catalyst was added to the reaction mixture at the desired temperature. A sample of 1 mL was taken every 30 min and centrifuged to separate the catalyst from methanol and oil to stop reaction.

2.4. Sample analysis

The resulting solution was then analysed by gas chromatography from Shimadzu GC-2010 Plus with DB-WAX capillary column and detected by flame ionization detector (FID). Methyl valerate was used as an internal standard. The column temperature was programmed as follows: starting at 40 °C with a holding time of 3 min, ramping to 235 °C at a rate of 15 °C/min, holding for 2 min, ramping to 260 °C at a rate 10 °C/min, and holding for 20 min. The weight fraction is calculated by the following eq. (3);

$$\text{Weight fraction} = \frac{C_{t,i} \times MW \times V_{t,0}}{Mass_{t,0}} \quad (3)$$

where $C_{t,i}$ is concentration of oleic acid at a different time, MW is molecular weight of oleic acid, $V_{t,0}$ is initial reaction volume of resulting sample and $Mass_{t,0}$ is initial total mass of reaction mixture.

3. Simulation

Simulations were carried out under the steady state condition using the AspenPlus® software version 8.0. The RADFRAC model has been used in the design of RD columns [7]. RADFRAC relies on an equilibrium stage model. Besides, pure component properties and Dortmund-UNIFAC interaction parameters, the column heat loss and the reaction kinetics data were incorporated into the simulator. The number of stages was counted from the top to the bottom, with the first stage is the condenser and the stage N is the reboiler. Oleic acid and triolein were used as a model compound for FFA and triglyceride, respectively.

3.1. Validation

3.1.1 Kinetic model of esterification. Figure 3 shows the experimental results in the dot and simulation results in continuous lines obtained from the proposed kinetic rate model [6]. The results indicated that there was no significant difference in the experimental results (esterification of oleic acid and methanol) and simulation result of esterification of decanoic acid and methanol provided by Steinigeweg and Gmehling [6].

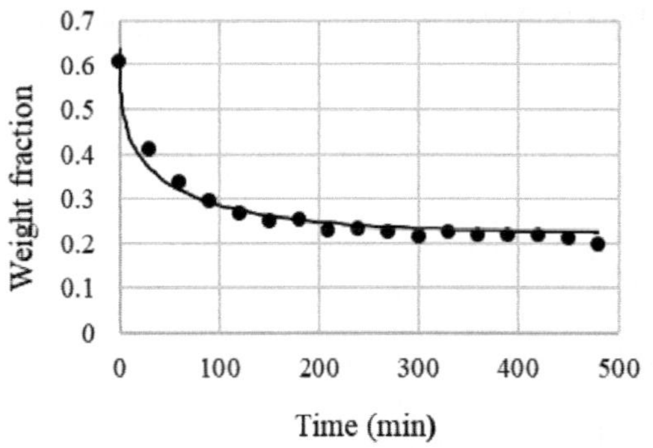

Figure 3. Kinetic model fitted eq. 1. (continuous line) and experiment (•) condition are T=60 °C methanol to oil ratio 5:1, X^0_{water} =0.019.

The kinetic constants and activation energies were reported by Steinigeweg and Gmehling [6], as presented in table 2, for the esterification of decanoic acid using Amberlyst-15. However, oleic acid is used as the feedstock in this work; therefore, the validation of FFA conversion as a function of reflux ratio was also performed as shown in figure 4. It is found that the obtained simulation calculations are close to experimental results at various reflux ratios with the maximum error of 16%.

Table 2. Kinetic parameters of LHHW model (K_{sorp}= 2.766). [6]

Reaction	i	k_i^0 (mol /g s)	$E_{A,i}$ (kJ /mol)
Esterification	1	3.1819×10^6	72.23
Hydrolysis	-1	3.5505×10^5	71.90

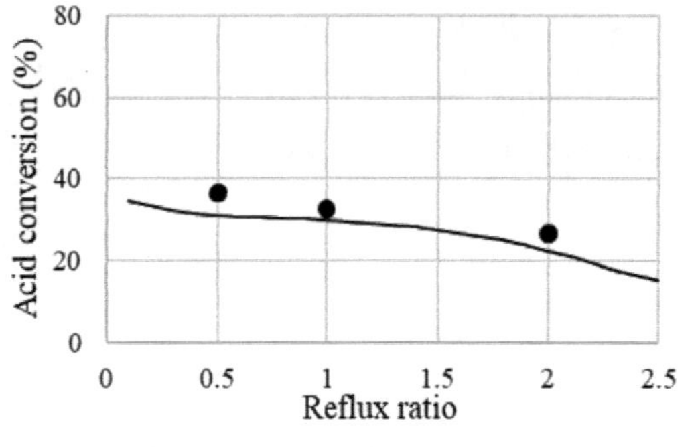

Figure 4. Acid conversion as a function of reflux ratio. Experimental data [6] (•) and simulation results (continuous line).

3.1.2 Transesterification. The kinetic constants were reported by Pasupulety et al [10], presented in table 3, for the transesterification of soybean oil catalysed by $CaO/n\text{-}Al_2O_3$ (neutral- Al_2O_3) and the value of activation energy was found to be 30.7 kJ/mol. However, triolein is used as feedstock in this work; therefore, the validation of FAME yield as a function of time was also performed as shown in figure 5. It was found that the obtained simulation calculations are close to the experimental results at different time with the maximum error of 27%. Although this seems to be excessive but the equilibrium are similar, whereby, the simulation relies only on the equilibrium.

Table 3 Pseudo-first order reaction rate constants at different temperatures. [10]

Temperature (°C)	Rate constant, k (1/min)	R^2
125	0.0029	0.99
150	0.0047	0.997
175	0.0072	0.994
200	0.013	0.999

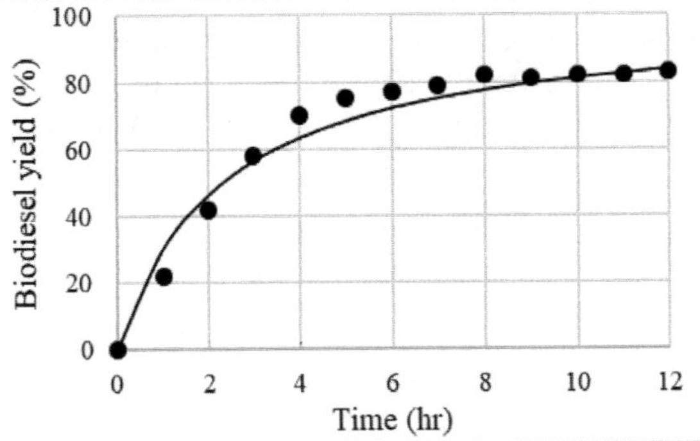

Figure 5. FAME yield as function of time. Experimental data [10] (•) and simulation result (continuous line) T = 150 °C, methanol to oil = 9:1 at pressure = 1 bar.

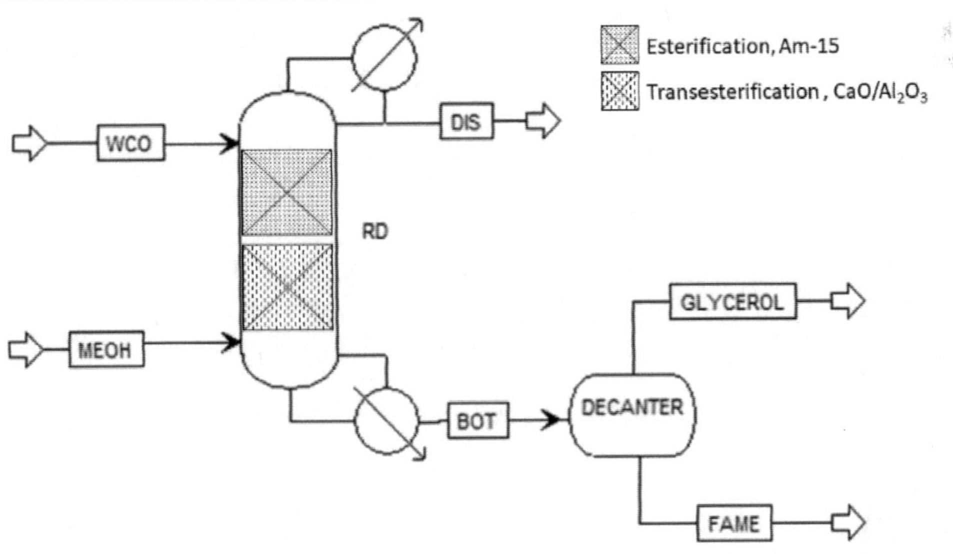

Figure 6. Ester-transesterification in single reactive distillation for biodiesel production.

3.2 Ester-transesterification in a single reactive distillation

Petchsoongsakul et al [7] proposed the ester-transesterification in a single RD. The upper column was packed with Amberlyst-15 for esterification and the lower column was packed with CaO/Al_2O_3 for transesterification. Methanol and waste cooking oil were co-fed at the second stage.

In this work (Figure 6), the same concept has been adopted but the feedstock contains water and the kinetic model of the esterification has been changed to LHHW model in order to account the influence of water on the operating parameter of the reactive distillation.

3.2.1 Number of stages. To reduce FFA below 1 wt%, 5 reactive stages were required for the esterification and 9 stages of transesterification to produce biodiesel according to EN standard. The obtained biodiesel yield was higher than 96.5% using the methanol to oil ratio of 6:1, the reflux ratio of 0.1 and the reboiler duty in the range of 100-150 kW as depended on the water content in WCO. WCO contaminated with water content of 2-8 wt% and FFA of 10 wt% as the oil feedstock. The results are shown in figure 7. It was found that the biodiesel purity was increased with the increase of water content from 2 to 8 wt% while biodiesel yield was no significant different when the water content in WCO was increased. However, the reaction rate of transesterification was decreased with increasing amount of water resulting in higher biodiesel purity when using higher amount of water in WCO. Figure 8 shows the effect of the water content in feedstock on the number of stage. It was found that the water content does not show significant effect on the number of stages because the water as well as methanol were removed to the top of the column resulting to a trace amount of the water was in biodiesel at the transesterification section in the bottom of column. The water content at this reaction section was less than 0.1 wt%.

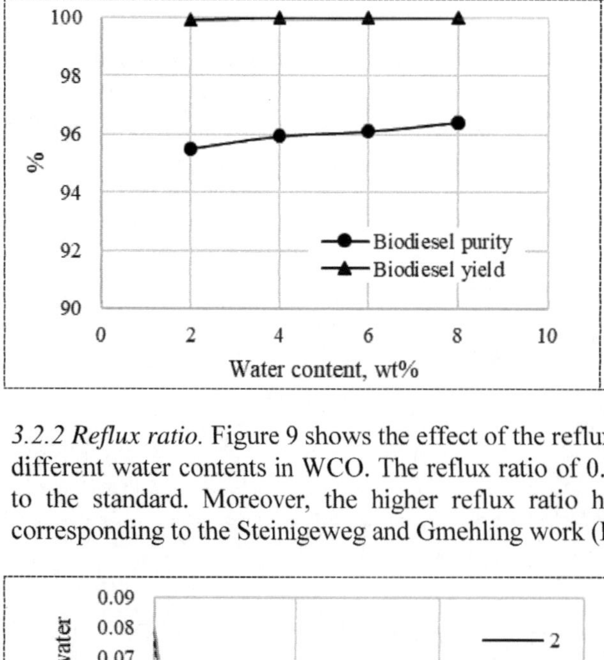

Figure 7. Biodiesel yield and purity as function of water content in feedstock. Condition are methanol to oil ratio 6:1, reflux ratio of 0.1 and reboiler duty of 100 kW.

3.2.2 Reflux ratio. Figure 9 shows the effect of the reflux ratio on the purity of produced biodiesel with different water contents in WCO. The reflux ratio of 0.1 was used to purify the biodiesel conforming to the standard. Moreover, the higher reflux ratio has a negative effect on the esterification as corresponding to the Steinigeweg and Gmehling work (Figure 4) [6].

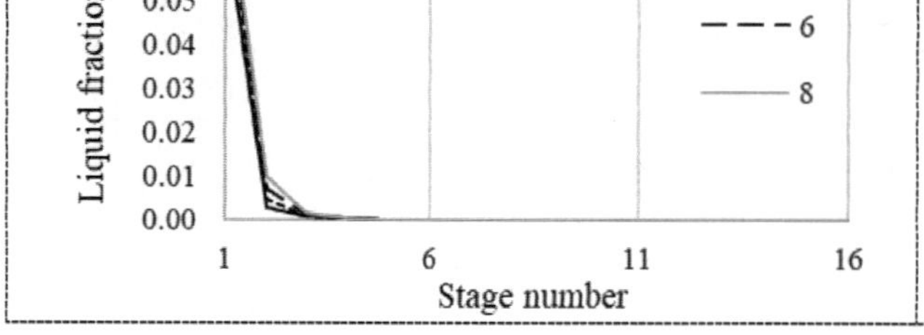

Figure 8. Water mass fraction in reaction section different water content in feedstock 2-8 wt%.

3.2.3 Methanol feed location. Liquid mass fractions of the components are presented in figure. 10A and B. Where, in 10A, methanol is fed at the middle of column and high liquid fraction of product (FAME) is achieved. On the other hand, in figure 10B, methanol is fed at the bottom of the column and the product's liquid fraction is lowered. In the current work, the bottom of the column has high temperature, resulting to sudden vaporization of methanol as found in the lower level of methanol in the liquid phase. This is leading to lower reaction rate which confirm by the biodiesel yield and purity as indicated in figure 11. The biodiesel yield and purity become lower when methanol was fed at the lower section of column. This was not corresponding to the conventional biodiesel production in RD. The oil feedstock is usually fed at the top of the column; while methanol is fed at the bottom [6, 7].

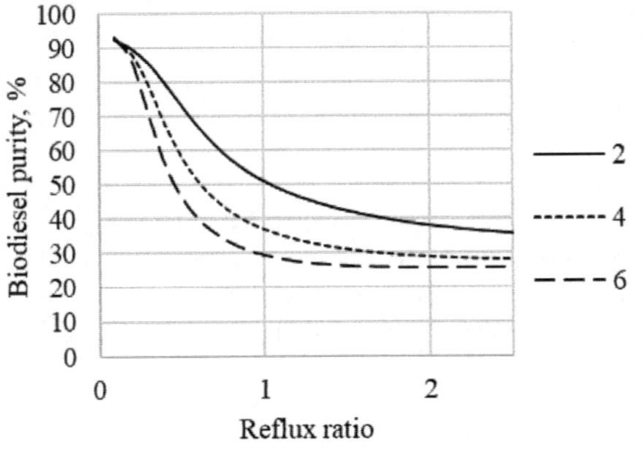

Figure 9. Biodiesel purity as a function of reflux ratio different water content (2-6 wt%) in oil feedstock. Conditions are methanol to oil ratio 6:1 and reboiler duty of 100kW.

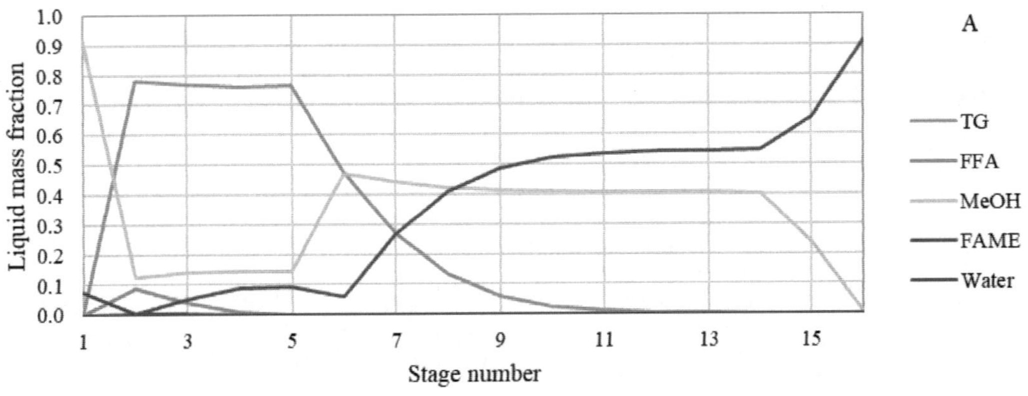

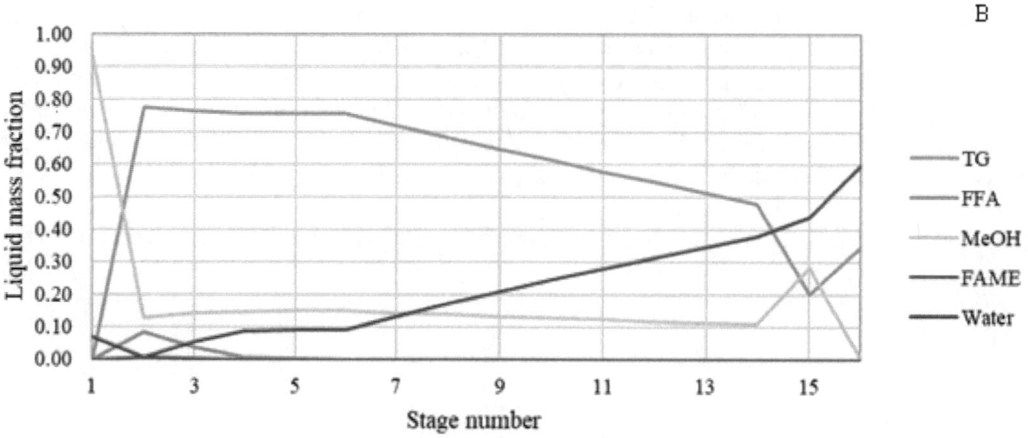

Figure 10. Liquid mass fraction in ester-transesterification in a single reactive distillation feed oil at stage 2 and (a) feed methanol at stage 6 (first stage for transesterification) and (b) feed methanol at the bottom of column using water content 4 wt%, reflux ratio of 0.1 and reboiler duty of 100 kW. (TG = triolein, FFA = oleic acid, MeOH = methanol and FAME = fatty acid methyl ester).

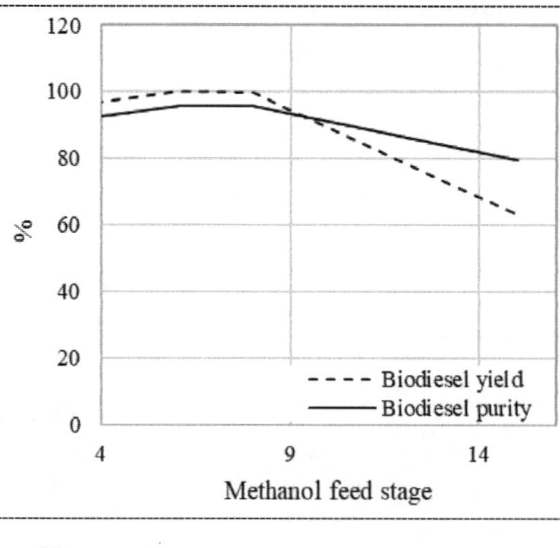

Figure 11. Function of biodiesel purity and yield methanol feed stage using water content of 4 wt% in WCO.

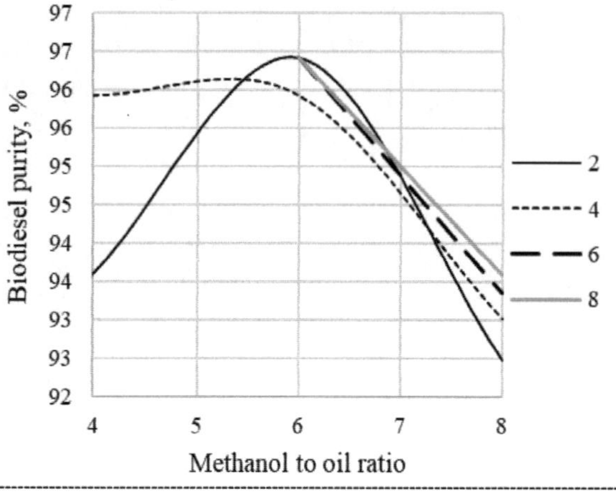

Figure 12. Biodiesel purity as function of methanol to oil ratio using different water content (2-8 wt%) of WCO.

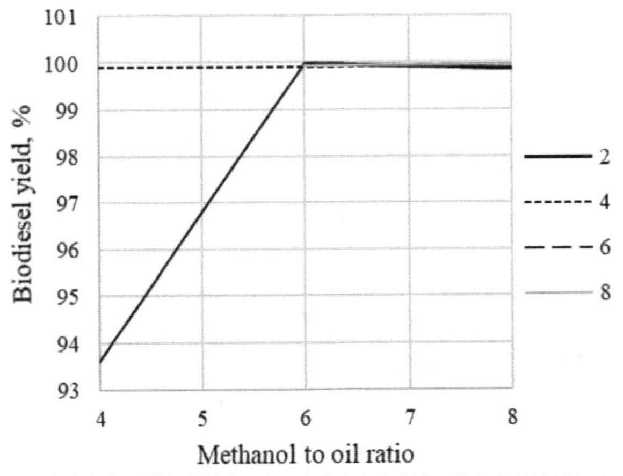

Figure 13. Biodiesel yield as a function of methanol to oil ratio using different water content (2-8 wt%) in WCO.

3.2.4 Methanol to oil feed ratio. Figure 12 and 13 show the biodiesel purity and yield respectively, as function methanol to oil mole ratio using different water content (2-8 wt%) in WCO. The optimum ratio was 6:1 because the excess methanol excess was contaminated in biodiesel at the bottom product.

4. Conclusions

The kinetic model of the decanoic acid reaction can represent the oleic acid esterification using Amberlyst-15 as solid acid catalyst. Transesterification, on the other hand, was used Pseudo-homogeneous kinetic model because of the trace amount of water was in this reaction zone. Therefore, the LHHW has been adopted to demonstrate the effect of the contaminated water on the esterification of WCO. The simulation results of the single reactive distillation for simultaneous ester-transesterification to produce biodiesel from WCO has found the optimum design of the RD column with 16 stages, consisting of 5 esterification stages and 9 transesterification stages. The methanol to oil ratio was kept for 6:1, when WCO was fed at top of the column and methanol was fed at the first stage of transesterification using the reflux ratio of 0.1 and the reboiler duty in the range of 100-150 kW. Biodiesel yield and purity more than 96.5% were obtained using WCO with contaminated water content of 2-8 wt% and FFA 10 wt%. This design of a single reactive distillation has a capability to withhold the water contaminated in the range of 2 to 8 wt% for simultaneous ester-transesterification of WCO.

6. References

[1] G. Suyin, Ng.H. Kiat, C.P. Hinn, L.F. Lim 2012 Heterogeneous free fatty acids esterification in waste cookingoil using ion exchange resins. *Fuel Processing Technology* **102** 67-72

[2] Son, S. Mo, K. Katsuk 2011 Transesterification of sunflower oil in a countercurrent trickle-bed reactor packed with a CaO catalyst. *Chemical Engineering and Processing: Process Intensification* **50(7)** 650-654

[3] L. Zih-Hua, L. Pei-Hsuan, W. Jeffrey C.S, H. Yu-Tzu, L. Kuen-Song, W. Kevin C. W 2013 A stirring packed-bed reactor to enhance the esterification–transesterification in biodiesel production by lowering mass-transfer resistance. *Chemical Engineering Journal* **234** 9-15

[4] B. Mauro, G. Giuseppe 2018 A Simple Pseudo-Homogeneous Reversible Kinetic Model for the Esterification of Different Fatty Acids with Methanol in the Presence of Amberlyst-15. **Vol. 11**, 1843

[5] Dj. Vujicic, D. Comic, A. Zarubica, R. Micic, G. Boskovic 2003 Kinetics of biodiesel synthesis from sunflower oil over CaO heterogeneous catalyst. *Fuel* **89(8)** 2054-2061

[6] S. Sven, G. Jürgen 2003 Esterification of a fatty acid by reactive distillation. *Industrial & Engineering Chemistry Research* **42(15)** 3612-3619

[7] N. Petchsoongsakul, K. Ngaosuwan, W. Kiatkittipong, F. Aiouache, S. Assabumrungrat 2017 Process design of biodiesel production: Hybridization of ester-and transesterification in a single reactive distillation. *Energy Conversion and Management* **153** 493-503
[8] L. Xuejun, H. Huayang, W. Yujun, Z. Shenlin, P. Xianglan 2008 Transesterification of soybean oil to biodiesel using CaO as a solid base catalyst. *Fuel* **87(2)** 216-221
[9] C. Guan-Yi, S. Rui, Y. Bei-Bei, S. Jia-Fu, L. Shang-Yao, L. Chang-Ye 2016 Remarkably enhancing the biodiesel yield from palm oil upon abalone shell-derived CaO catalysts treated by ethanol.*Fuel Processing Technology* **143** 110-117
[10] P. Nagaraju, G. Kamalakar, L. Yuanqing, R. Garry L, Ng. Flora TT 2013 Production of biodiesel from soybean oil on CaO/Al2O3 solid base catalysts. *Applied Catalysis A: General* **452** 189-202

Acknowledgement
The authors would like to acknowledge the support from the Research Chair Grant under the National Science and Technology Development Agency.

IOP Conf. Series: Materials Science and Engineering **559** (2019) 012015 doi:10.1088/1757-899X/559/1/012015

Impact of Oxygen Vacancy on the Photocatalytic Selective Hydrogenation of 3-Nitrostyrene via Calcination of TiO_2

Saknarin Chaitaworn[1,a], Okorn Mekasuwandumrong[2] and Piyasan Praserthdam[1*]

[1] Center of Excellence on Catalysis and Catalytic Reaction Engineering, Department of Chemical Engineering, Faculty of Engineering, Chulalongkorn University, Bangkok 10330, Thailand
[2] Department of Chemical Engineering, Faculty of Engineering and Industrial Technology, Silpakorn University, Nakhon Pathom 73000, Thailand

[a] Corresponding author: piyasan.p@chula.ac.th

Abstract. In this work, P25-TiO_2 was treated by calcination in air at difference temperature (600-900°C) for 5h. Treated catalysts were tested in photocatalytic selective hydrogenation of 3-nitrostyrene (3-NS) to 3-vinylaniline (3-VA). The properties of treated catalysts were characterized by using XRD, XPS, and UV-Vis spectroscopy. 3-vinylaniline and acetone were produced from photocatalytic (λ > 300nm) selective hydrogenation reaction of 3-nitrostyrene in 2-propanol and suspension with P25-TiO_2. The Ti^{3+} at locate with oxygen defect, which at below conduction band were acted as active site for trapped electron to conduction band. These effects leading to excess oxygen defect caused low activity due to recombination center. An optimal calcination temperature was 700°C ascribed to the enhancement of high conversion (71%) with 100% 3-VA selectivity due to optimal oxygen vacancies.

1. Introduction

Aniline is organic molecules and fine chemical, in which important intermediate for the synthesis of pharmaceutical, polymers, and herbicides. Functionalized aniline is generally produced by hydrogenation of nitroaromatic [1, 2]. The most catalyst hydrogenation for nitroaromatic compounds of the other reducible group (e.g. vinyl group and carbonyl group) in active with oxides catalyst or noble metal oxides catalyst and presence with molecular hydrogen (H_2) [1]. For the alternative produce to aniline from transformation of nitroaromatic is photocatalytic process include photooxidation and photoreduction without hydrogen gas. This process produces 3-vinylaniline from 3-nitrostyrene by photoreduction [1, 2].

Photocatalytic hydrogenation on semiconductor particles has several unique comparing with catalytic hydrogenation on metals or homogeneous catalysis using metal complexes. It can be defined as a light-driven reaction at solid/liquid or solid/gas interface [3]. Titanium dioxide (TiO_2) has three common polymorphic forms such as anatase, rutile, brookite [4]. TiO_2 was activated by irradiation by ultraviolet (UV) light, in which leads to enhance for selective hydrogenation of 3-NS. TiO_2 photocatalytic has widely been used in many applications such as degradation of pollutants, water splitting, and selective transformation of reactant to targeted products [5]. Shiraishi et al. [1] studied the photocatalytic selective hydrogenation of nitroaromatics on TiO_2 at room temperature and atmospheric pressure leads to the high yield of 3-VA, furthermore, noble metal dopants (Pt, Pd, Ag,

Content from this work may be used under the terms of the Creative Commons Attribution 3.0 licence. Any further distribution of this work must maintain attribution to the author(s) and the title of the work, journal citation and DOI.
Published under licence by IOP Publishing Ltd

Au) showed much lower yield due to coverage of active site and suppression of nitroaromatic adsorption. Imamura et al. [2] studied the selective photocatalytic hydrogenation of 3-NS and found that the other sites products including 3-nitroethybenzene (3-ENB) and 3-aminoethylbenzene (3-EA) were not formed. The high yield of 3-VA demonstrated that neither reaction benzene ring reduction of vinyl group nor re-oxidation of the amino group of 3-VA occurred in the system, however, stoichiometry generate electron and hole were selectively used for only reduction 3-NS to 3-VA and oxidation was produced ketone group from alcohol [2].

In this work, the catalytic performance of P25-TiO_2 calcination at difference temperature in photocatalytic selective hydrogenation of 3-NS were investigated. The physiochemical properties of treated catalysts were also determined by XRD, XPS, UV-Vis.

For mechanism photocatalytic hydrogenation selective 3-NS, when TiO_2 was irradiated by UV light, charge separation occurs formed positive hole in the valence band and formed electrons in conduction band [6]. This reactions P25-TiO_2 suspended in alcohol solvents as a hydrogen source containing with nitroaromatic under inert gas atmosphere and P25-TiO_2 (anatase 83% rutile 17%) used for semiconductor and non-metal dopant. The positive hole oxidize with alcohol and formed oxidation produce ketone and protons and nitroaromatic reduced with electrons and protons formed reduction produce aniline [1, 2].

2. Experimental

2.1. Catalyst preparation

Commercial Degussa P25-TiO_2 was used as the catalyst 2g then P25-TiO_2 was transferred into the crucible and followed by the calcination at various temperatures (600, 700, 800, 900°C) in a muffle furnace for 5h.

2.2. Photocatalytic test

The mixture 200micromoles of 3-nitrostyrene, 20ml of isopropanol and 0.2g of catalyst was transferred into 75ml pyrex glass tube reactor. The mixture was sonicated for 5min then the solution was purged with N_2 for 5min. The pyrex glass tube containing substrate, solvent and catalyst were photoirradiated under mercury lamp Philip 500w ($\lambda > 300$nm) for 6h with magnetic stirrer. After that the reaction catalyst was removed by filtration and the resulting products was analyzed by GC-FID (Shimadzu with Rtx®5 columns and Shimadzu DB1-wax columns).

3. Results and discussion

3.1. Characterization of P25-TiO_2 catalysts after calcination at difference temperature

3.1.1. X-ray diffraction (XRD)

The XRD patterns of P25-TiO_2 were treated with difference calcination temperature are shown in Figure 1. The P25-TiO_2 catalyst exhibited the characteristic peaks for anatase (101) phase at 2Θ degrees = 25.3° and rutile (110) phase at 2Θ degrees = 27.4° [7]. In this work, P25-TiO_2 was treated at calcination temperature range of 600-900°C for 5h. P25-TiO_2 including both anatase and rutile phases. Calcination at 600°C the peaks of anatase decreased slightly along with increased rutile. Moreover, higher temperature calcination leads to the rapid increased rutile phase and crystallite size [8, 9]. P25-TiO_2 calcination at 700°C, peaks of anatase phase component decreased rapidly and 77% of rutile phase was formed. Anatase phase was decreased and lastly disappeared after calcination at 800 and 900°C. The calcination temperature clearly influenced the phase composition and crystallization. It is reasonable to suggest that high temperature calcination leads to phase transformation anatase to rutile [8, 9]. Therefore, phase transformation is associated with crystal growth. Table 1 shows the phase composition and crystallite size of P25-TiO_2 calcination at temperature range of 600-900°C.

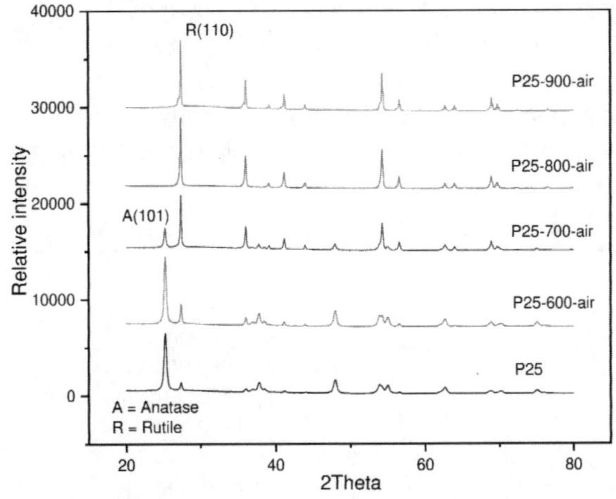

Figure 1. XRD patterns of P25-TiO_2 samples treated with difference calcination temperature.

Table 1. Phase composition and crystallite size of P25-TiO_2 calcination difference temperature.

Sample	%Anatase	%Rutile	Crystallite size Anatase (101) (nm)
P25	83	17	23
P25-600-air	73	27	29
P25-700-air	23	77	43
P25-800-air	-	100	n/a
P25-900-air	-	100	n/a

3.1.2. X-ray photoelectron (XPS)

XPS was used for determine and analyze the surface chemistry elemental composition, chemical state of the TiO_2 [10]. The deconvolution peaks of Ti 2p core levels of pure TiO_2 was fitted with four peaks consistent with titanium dioxide (Ti^{4+}) and titanium sub oxide (Ti^{3+}) are shown in Figure 2 [10].

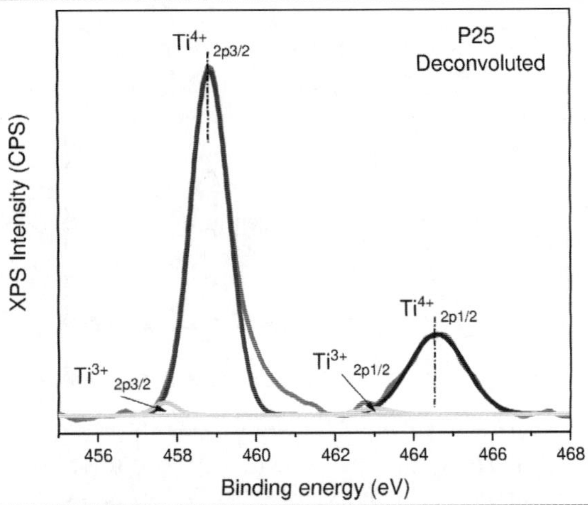

Figure 2. XPS Ti 2p region spectra with deconvolution peaks with Gaussian fitting curve of P25-TiO_2 samples.

However, TiO_2 having typical Ti 2p binding energies both Ti $2p_{3/2}$ and Ti $2p_{1/2}$. At peak 458.5 eV and 464.5eV are associated with the Ti^{4+} of Ti $2p_{3/2}$ and Ti $2p_{1/2}$ valence state respectively [10-12]. Titanium sub oxide has two peaks at 457.7 and 463.0eV are associated Ti^{3+} of Ti $2p_{3/2}$ and Ti $2p_{1/2}$ valence state respectively, which corresponding with oxygen vacancies (O_V) defect in the system [11, 14]. For deconvolution peaks of Ti 2p core levels with Gaussian fitting are shown in Figure 2 including both Ti $2p_{3/2}$ and Ti$2p_{1/2}$.

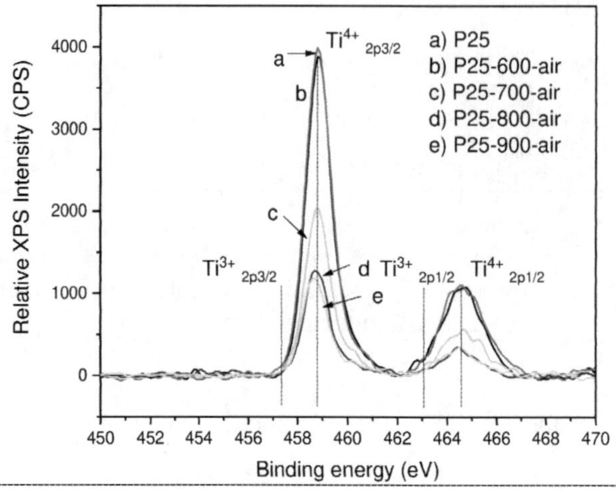

Figure 3. XPS Ti 2p region spectra with difference temperature calcination of P25-TiO_2 samples.

The relative concentration Ti^{3+} of P25-TiO_2 was increased when the high calcination temperature, the concentration of surface Ti^{4+} was reduced along with a loss of oxygen from the surface TiO_2, Ti^{4+} ions receive electrons from lattice oxygen, which are removed from stoichiometric TiO_2 are shown in Figure 3 [13]. Meanwhile, the concentration Ti^{3+} was decreased after calcination at 900°C due to more energy was supplied electron became very weak due to the phase transition reaction from anatase to rutile [13].

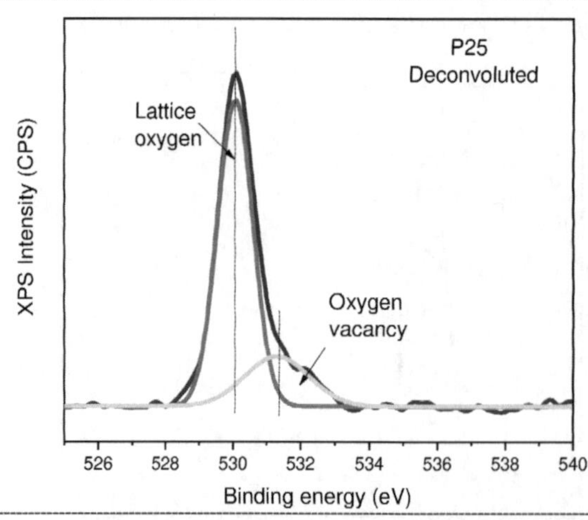

Figure 4. XPS O1s region spectra with deconvolution peaks with Gaussian fitting curve of P25-TiO_2 samples.

The deconvolution peaks of O1s core levels of P25-TiO_2 are shown in Figure 4. It showed two main peaks centered at 529.80 and 531.30 eV, which are assigned to Ti^{4+} and Ti^{3+} - oxygen vacancies sites, respectively [11-12, 14]. In addition P25-TiO_2 was calcination at high temperature, the peak at 529.80 was reduced due to broken Ti-O-Ti, then oxygen vacancies was formed and corresponding with the calcination in oxygen deficient environmental are shown in Figure 5 [12].

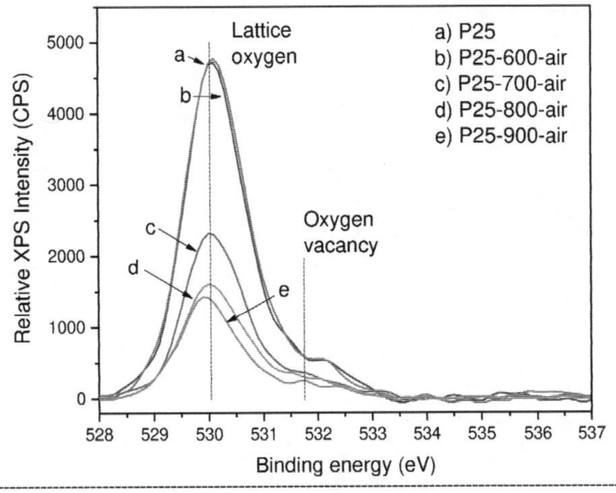

Figure 5. XPS O1s region spectra with difference temperature calcination of P25-TiO_2 samples.

3.1.3. UV-Visible spectroscopy (UV-Vis)

The influences of calcination temperature on the light absorption were investigated by UV-Vis spectroscopy. The UV-Vis spectra of P25-TiO_2 calcination at difference temperatures are shown in Figure 6. Increasing of calcination temperature leads to the adsorption of visible wavelength, which could be ascribed to the absorption of light caused by excitation of electrons from valence band to conduction band. The difference of adsorption wavelength was changed due to change of crystallite size and phase structure [8].

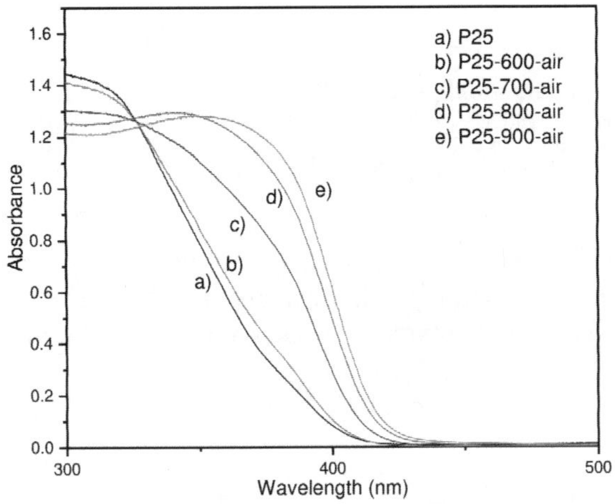

Figure 6. UV-Vis wavelength absorption of P25-TiO_2 samples with difference temperature calcination.

The adsorption edges wavelength after calcination at 600°C indicated the decrease band gap energy as shown in Figure 7. The band gap energy can be determined from a plot of $(h\nu\alpha)^{1/2}$ versus photon energy ($h\nu$). Calcination of P25-TiO_2 at high temperature leads to decrease of band gap energy because the increase of crystallite size results in the decrease of band gap. On the other hand, after calcination at 600°C might be a result of phase transformation from anatase to rutile [8].

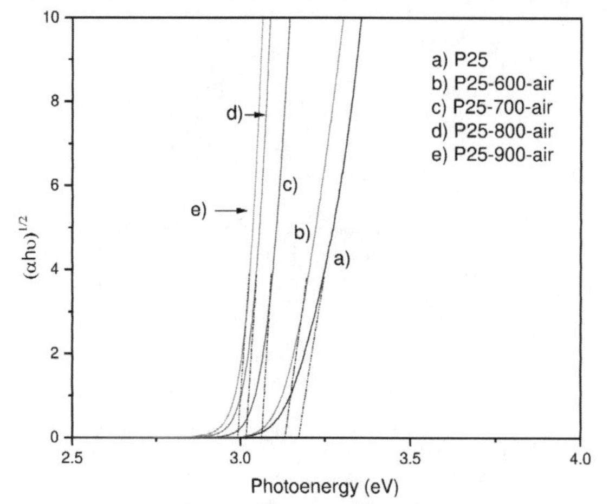

Figure 7. Band gap that plot between $(h\nu\alpha)^{1/2}$ versus photon energy of P25-TiO_2 with difference calcination temperature.

Table 2. UV-Vis results and band gap energy of all catalysts.

Sample	Wavelength (nm)	Band gap (eV)
P25	392	3.18
P25-600-air	405	3.14
P25-700-air	416	3.05
P25-800-air	417	3.01
P25-900-air	418	2.98

The adsorption edges wavelength and band gap are shown in Table 2. The adsorption edge wavelength were increased due to phase transition anatase to rutile. Furthermore band gap were decreased due to increasing of crystallite size.

3.2. Photocatalytic activity
The photocatalytic activity of the P25-TiO_2 after calcination at difference temperature was carried out by photocatalytic selective hydrogenation 3-NS, then photocatalytic (λ > 300nm) irradiated 6h selective hydrogenation reaction of 3-nitrostyrene 200micromoles in 2-propanol 20ml and suspension with P25-TiO_2 0.2g

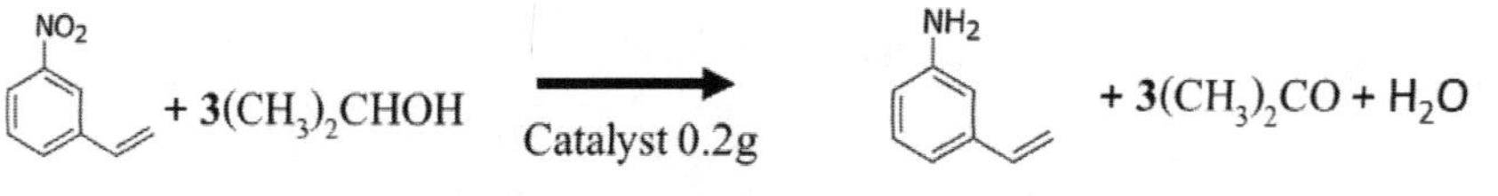

Table 3. Results of photocatalytic hydrogenation selective 3-NS over difference temperature calcination P25-TiO_2 by UV light irradiated 6h.

Sample	%Conversion of 3-NS	%Selectivity to 3-VA
P25	45.4	96.7
P25-600-air	43.1	100
P25-700-air	70.9	100
P25-800-air	57.8	100
P25-900-air	51.5	100

The %conversion 3-NS and %selective 3-VA are shown in Table 3, which show the catalytic performance of P25-TiO_2 calcination at difference temperature. P25-TiO_2 calcination at 700°C shows the highest conversion 3-NS with nearly 100% 3-VA selectivity. The enhancement of photocatalytic activity can be attributed to the decreasing of band gap energy, including of oxygen vacancies contents and suitable phase components. It is well known that mixture of anatase and rutile phase have efficiency photoactivity [8]. It also reported that in photocatalytic hydrogenation selective 3-NS the surface Ti^{3+} atoms of rutile TiO_2 behaved as the absorption site for nitroaromatic trapping for photoformed at conduction band [1, 2]. In our present work, P25-TiO_2 exhibit insufficient photoactivity 3-NS ,so improvement photoactivity the calcination at high temperature leads to transition of morphology, band gap, oxygen vacancy and phase components of P25-TiO_2 results in higher photoactivity. Moreover, P25-TiO_2 calcination at 700°C with relative mixed phase ratio (rutile 77%, anatase 23%) exhibit highest photoactivity due to the including of oxygen vacancy. On the other hand, P25-TiO_2 calcination above 800°C leads to excess oxygen vacancy might be caused as the recombination center due to un-localized below at conduction band, which blocked the electron transferring to conduction band [9]. On the other hand, mixture of anatase and rutile phase, moreover increasing rutile content attributed to enhancement photoactivity due to optimal oxygen vacancies leads to inhibit recombination results in higher photoactivity due to inhibit recombination electron-hole. Thus both oxygen vacancy and interface between anatase and rutile has a positive role play photocatalytic [10].

4. Conclusion

Calcination treatment of P25-TiO_2 influence on microstructure such as crystallite size, wavelength absorption, rutile phase content, oxygen vacancy sites and photocatalytic activity. P25-TiO_2 treated at optimal calcination temperature at 700°C exhibit the highest photoactivity due to the including of oxygen vacancy, Ti^{3+} sites and interface between anatase and rutile, which improve the nitroaromatic trapping and retarded electron-hole recombination process. However, excess oxygen vacancy leads to recombination center results in the decreasing of photocatalytic activity.

5. Reference

[1] Shiraishi Y, Togawa Y, Tsukamoto D, Tanaka S and Hirai T 2012 *ACS Catalysis* **2** 475-2481
[2] Imamura K, Yoshikawa T, Hashimoto K and Kominami H 2013 *Applied Catalysis B: Environmental* **134-135** 193-197
[3] Berguerand C *Industrial & Engineering Chemistry Research* 2015 **54** 8659-8669
[4] Reyes-Coronado D, Rodriguez-Gattorno G, Espinosa-Pesqueira M. E, Cab C, Coss R de and Oskam G *Nanotechnology* 2008 **19** 145605
[5] Fujishima and Honda K *Nature Article* 1972 **238** 37-38
[6] Munir S, Dionysiou D. D, Khan S.B, Shah S. M, Adhikari B and Shah J A *Photochem Photobiol B* 2015 **148** 209-222
[7] Ohtani B, Prieto-Mahaney O. O, Li D and Abe R *Journal of Photochemistry and Photobiology A: Chemistry* 2010 **216** 179-182
[8] Wang G, Xu L, Zhang J, Yin T and Han D *International Journal of Photoenergy* 2012 1-9
[9] Fernandes Machado N.R.C and Santana V.S *Catalysis Today* 2005 **107-108** 595-601

[10] Bharti B, Kumar S, Lee H.N and Kumar R. *Sci Rep* 2016 **6** 32355
[11] Tshabalala Z. P, Motaung D. E, Mhlongo G. H and Ntwaeaborwa O. M *Sensors and Actuators B: Chemical* 2016 **224** 841-856
[12] Paul K. K, Jana S and Giri P. K *Particle & Particle Systems Characterization* 2018 **35**
[13] Xiong L -B, Li J -L, Yang B and Yu Y *Journal of Nanomaterials* 2012 1-13
[14] Wang X, Li Y, Liu X, Gao S, Huang B and Dai Y *Chinese Journal of Catalysis* 2015 **36** 389-399

Acknowledgments
This research is supported by the Ratchadapisek Sompoch Endowment Fund (2016), Chulalongkorn University (CU-59-006-IC) and Newton Mobility Grants.

A Comparative Study of the Divalent Transition Metal Oxide Supported on Magnesium Oxide Catalyst for Lactic Acid Production from Glucose

T Udomcharoensab[1] and P Praserthdam[1,a]

[1] Center of Excellence on Catalysis and Catalytic Reaction Engineering, Department of Chemical Engineering, Faculty of Engineering, Chulalongkorn University, Phyathai Road,Patumwan, Bangkok, 10330, Thailand

[a] Corresponding author: piyasan.p@chula.ac.th

Abstract. In order to achieve the lactic acid production of lower energy consumption, reactor corrosion rate, while being environmentally friendly and sustaining high catalytic activity of various divalent transition metal oxide (Cu, Co, Ni and Zn) supported on magnesium oxide having the hexadecyltrimethylammonium bromide (CTAB) as a capping agent was investigated, where the reaction was performed under mild conditions in an alkali additive-free environment. Among all of the divalent candidates, the ZnCTAB/MgO exhibits the highest activity with a glucose conversion of 64% at a 100°C and a lactic acid yield of 12% at 140°C, a pressure of 0.4 MPa, over 1 h of reaction time. This is due to the high basicity produced via the water chemisorption on such catalyst's surface.

1. Introduction

Over the last decade, global warming and energy crisis are major concerns caused by the increase in atmospheric carbon dioxide content and the depletion of fossil fuels. These issues force the chemical industry to focus more on the necessity to change from the feedstock of fossil fuels based to renewable resources.

Biomass is one of the suitable renewable carbon resources because it contains adequate carbon-carbon bonds to produce useful chemicals. Moreover, it does not increase the carbon dioxide in the atmosphere, making it a cleaner choice than the energy produced from fossil-based [1]. The major component of lignocellulosic biomass (wood, straw, and grasses) is the C6-sugars (glucose, fructose, and mannose), which are interesting precursors for the use in a wide range of high-value chemicals production.

The lactic acid (2-hydroxypropionic acid, $C_3H_6O_3$) is an important natural organic acid, which has broad applications in food, pharmaceutical, and cosmetic industries and it is also a building block for the synthesis of many useful chemicals, e.g., pyruvic acid, acetaldehyde, acrylic acid, and propanoic acid [2]. In recent years, lactic acid has gained more attention as a major feedstock for the manufacture of biodegradable plastics [poly(lactic acid), PLA], which can be decomposed by bacteria, while exhibiting a good mechanical strength. Thus, it is a suitable choice for the replacement of petroleum-based polymers reducing the environmental impacts [3].

Content from this work may be used under the terms of the Creative Commons Attribution 3.0 licence. Any further distribution of this work must maintain attribution to the author(s) and the title of the work, journal citation and DOI.
Published under licence by IOP Publishing Ltd

The lactic acid can be synthesized by two routes of either the fermentation of carbohydrates with microorganisms or the chemical synthesis. At present, over 90% of the industrial lactic acid production is by the fermentation of glucose from starch with enzyme or microbes. However, the biological process has some disadvantages, e.g., strict controls of reaction conditions, a long fermentation time, and difficult scale-up of the process. Hence, the chemical synthesis route is a more interesting choice [4].

Generally, in the chemical route, the lactic acid was synthesized via the hydrolysis of lactonitrile, formed by the reaction between acetaldehyde and hydrogen cyanide. This process generates a highly toxic cyanide compound that causes an environmental issue [5]. As a result, producing the lactic acid from biomass via the chemocatalytic systems has been proposed as a promising route.

It is well-known in a lactic acid production from sugar that the alkali is typically used as a catalyst. Various authors focused on the hydrothermal process with or without alkali solution and/or heterogeneous catalyst. They reported that the water in the subcritical and supercritical regions acts as an alkaline catalyst. Some researchers also studied the conversion of glucose into lactic acid at high temperature (above 250°C) for short time (60-90 s). In 2007, Yan *et al.* compared two alkaline catalysts including sodium hydroxide (NaOH) and calcium hydroxide [$Ca(OH)_2$] at 300°C for 60 s. The highest lactic acid yield of 27% was obtained for 2.5 M NaOH [6]. To enhance the production of carboxylic acids from carbohydrates, Wang *et al.* (2013) used a metal oxide as an oxidant catalyst. They compared the oxidative potential of different metal oxides (CuO, Fe_3O_4, Fe_2O_3, Al_2O_3, ZnO and TiO_2) and found that CuO gave the best activity with yields of lactic acid, acetic acid, and formic acid of 37.1%, 9.4%, and 4.9% respectively at 300°C for 60 s with CuO of 1.5 mmol and NaOH of 2.5 M [1, 7]. The previous study performed the reaction under a high concentration of alkali causing the corrosion to the reactor and pollute to the environment. To solve this problem, Huo *et al.* (2014) suggested that transition metal ions could improve lactic acid production from glucose and decrease the NaOH concentration at the same time. They compared various transition metal ions of Ni^{2+}, Zn^{2+}, Co^{2+}, Fe^{2+}, Fe^{3+}, Mg^{2+}, Cd^{2+}, Sn^{2+}, and Cu^{2+} obtained from metal chloride compounds. The highest yield of 25% for lactic acid was obtained with 0.01 M Ni^{2+} and 0.01 M NaOH at 300°C for 1 min [8]. In addition, there are several researches that used solid catalyst without the addition of an alkali solution. Xue and Gao investigated various types of feldspars (Albite and Potassium feldspar) and Bentonite (wet clay mineral) as a solid basic catalyst. The highest lactic acid yield from Albite feldspar was 12.37% at 260°C for 100 s, whereas Bentonite gave 11% yield of lactic acid at 275°C for 100 s [9, 10]. In 2016, their group also studied the possibility and effectivity of sodium silicate as a mild base catalyst. They compared various catalysts including Na_2SiO_3, MnO_2, TiO_2, $NaHCO_3$, and $Al(OH)_3$. From their result, the highest lactic acid yield (30%) was obtained with 0.6 mol/L Na_2SiO_3 at 300°C for 60 s [11].

Obviously, all previous studies were performed under extreme conditions, which has some disadvantage including high energy consumption, increase equipment investment and limitation for the scale-up. So, many studies at mild condition were carried out, i.e., lower temperature and longer time. In 2008, Onda *et al.* studied the effect of various metals (Ru, Cu, Ag, Pd, and Pt) supported on the activated carbon catalyst on the selectivity of lactic acid and gluconic acid from glucose. The Pt/C showed the highest activity with the lactic acid and gluconic yield of 47% and 46%, respectively at 80°C for 2 h under air bubbling [3]. They focused on another solid basic catalyst, the hydrotalcite. They investigated the effect of the calcination temperature of the catalyst. The result showed that the calcined hydrotalcite under air at 450°C gave the highest lactic acid yield (20%) at 50°C for 8 h with NaOH 50 mmol/L [12]. In 2015, Choudhary *et al.* used the magnesia supported-Cu oxide catalyst synthesized via a hydrothermal method, which makes the copper oxide species interacts strongly with the support. In addition, CTAB (hexadecyltrimethylammonium bromide) was used as the capping agent for such catalyst. They investigated the effect of calcination temperature on lactic acid production. From their result, CuCTAB/MgO calcined under air at 500°C gave the highest yield for lactic acid and exhibited the maximum number of Brønsted basic sites. The optimum condition for the reaction was 120oC for 1 h with 2.5 M of NaOH resulting in 70% of lactic acid yield. They reported

that CTAB favored the dispersion of metals on metal oxides and influenced the generation/regeneration of the active Cu species during calcination [13]. Li *et al.* (2017) studied the effect of common alkali metal hydroxides (NaOH and KOH) and alkaline-earth metal hydroxides ($Mg(OH)_2$ and $Ba(OH)_2$) on lactic acid yield. They found that $Ba(OH)_2$ exhibited higher catalytic activity than NaOH and KOH in the same OH^- concentration. They reported that the divalent Ba^{2+} ion played an important role in the conversion of glucose to lactic acid. The highest lactic acid yield of 78.3% was obtained at 25°C for 48 h under a nitrogen atmosphere.
From the previous research, most of all introduced reaction with the addition of alkali solution and many types of research showed the important role of the divalent ion of the catalyst in glucose conversion to the lactic acid. These inspired us to improve heterogeneous catalyst for glucose conversion to lactic acid without the addition of the alkaline solution, which reduces the reactor corrosion problem while being less toxic to the environment and the catalyst can be reused. So, this research focuses on the investigation of the effect of various divalent transition metal oxides supported on magnesium oxide catalyst using CTAB as the capping agent on lactic acid production from glucose performed under mild temperature and pressure in an alkali-free environment.

2. Experimental

2.1. Chemicals
The D-(+)-glucose, D-(-)-fructose, hexadecyltrimethy-ammonium bromide (CTAB), magnesium oxide (MgO), copper(II) nitrate trihydrate ($Cu(NO_3)_2{\bullet}3H_2O$), cobalt(II) nitrate hexahydrate ($Co(NO_3)_2{\bullet}6H_2O$), nickel(II) nitrate hexahydrate ($Ni(NO_3)_2{\bullet}6H_2O$), zinc acetate dihydrate ($Zn(CH_3COO)_2{\bullet}2H_2O$), lactic acid , formic acid and acetic acid used as reactants and/or as standard agents for the analysis were purchased from S.M. Chemical Supplies Co., Ltd.

2.2. Catalyst preparation
All catalysts were synthesized by the hydrothermal method described by Choudhary *et al.* with some modification [13]. First, the MgO was dispersed in deionized water by sonication. Then, an aqueous solution of 1mmol/g-catalyst of various metals [$Cu(NO_3)_2{\bullet}3H_2O$, $Co(NO_3)_2{\bullet}6H_2O$, $Ni(NO_3)_2{\bullet}6H_2O$ or $Zn(CH_3COO)_2{\bullet}2H_2O$] was added dropwise into the dispersed MgO solution under vigorous stirring. Thereafter, an aqueous CTAB solution of 1 mmol/g-catalyst was added into the mixture and vigorous stirring was continued for 3 h. The mixture was added in a 3-L autoclave prior to the heating to 180°C, where it was maintained at such temperature for 24 h. Afterward, it was slowly cooled down to room temperature. The obtained solid was washed with deionized water until the pH of the filtrate solution reaches 7.0 prior to another washing with ethanol. The solid was dried overnight at room temperature before being calcined in air at 500°C for 6 h with a ramp rate of 10°C/min. The prepared catalyst is characterized as a metal (Cu, Co, Ni, and Zn) catalyst supported on the MgO having CTAB as the capping agent.

2.3. Catalytic reaction procedure
The reaction testing was conducted in a Teflon-liner-fitted 50-mL-stainless-steel autoclave. First, 0.36 g of catalyst was load into the autoclave before the 0.54 g dissolved glucose dissolved in 30 mL of deionized (DI) water was added. Afterward, the autoclave was sealed, followed by the purging with nitrogen (N_2) three times at 0.4 MPa. After the purging, the autoclave was placed in a preheated oil bath at 140°C and stirred with a magnetic stirrer at 400 rpm and reacted at 140°C for 1 h. After the reaction, the autoclave was cooled down to room temperature in an ice-water bath immediately. The resulting solution was filtrated with a 0.2 μm syringe filter and analyzed by the high-performance liquid chromatography (HPLC).

2.4. Product analysis
The resulting solution was analyzed by HPLC, Agilent LC 1100 using an Agilent Hiplex-H column equipped with a UV-detector with a wavelength of 195 nm used in the detection of glucose and fructose [14]. In addition, the wavelength of 210 nm was used to detected organic acids, i.e., lactic acid, formic acid, and acetic acid. A dilute sulfuric acid aqueous solution of 5 mM in deionized water was used as the mobile phase at a flow rate of 0.6 mL/min. The conversion and yield were determined from the calibration curve.
The glucose conversion and yield of organic acid were calculated from the following equation:

$$\text{Conversion (\%)} = \frac{\text{Moles of introduced substrate - Moles of remained substrate}}{\text{Moles of introduced substrate}} \times 100 \quad (1)$$

$$\text{Product yield (\%)} = \frac{\text{Moles of product} \times \text{Number of carbon in product}}{\text{Moles of introduced substrate} \times \text{Number of carbon in substrate}} \times 100 \quad (2)$$

2.5. Catalyst characterization
The crystal structure and X-ray diffraction (XRD) patterns were analyzed by an X-ray diffractometer (Bruker D8 Advance) using Cu Kα irradiation at a 2θ angle between 10° and 80° with a step of 0.05°/s. The lattice parameter and d-spacing were calculated via the Bragg's law.
Temperature programmed reduction (TPR) was used to determine the reducibility of the catalyst. In H_2-TPR experiments, 0.08 g of catalyst was placed into the U-tube quartz reactor and pretreated under nitrogen (flow 25 ml/min) at 250°C for 1 h to remove water and impurities in the catalyst. After cooling down to 30°C, the catalyst was heated from room temperature to 650°C at a rate of 10°C/min under a flow of 10%H_2/Ar mixture (25 ml/min), in which the consumption of H_2 was recorded by the TCD.
The basicity of the catalyst was measured by carbon dioxide temperature programmed desorption (CO_2-TPD) and the titration with benzoic acid. For CO_2-TPD, each catalyst (0.1g) was charged into the U-tube quartz reactor. It was pretreated at 450°C with a temperature ramp of 10°C/min for 1 h under flow of Helium at 25 ml/min to remove any physisorbed organic molecules. Then, it was cooled down to 60°C followed by the saturation with CO_2 (flow at 25 ml/min) for 30 min. After which, the gas was switched to helium (25 mL/min) to remove the physisorbed CO_2. When the baseline is constant, the temperature of the furnace was increased to 750°C with a temperature ramp of 10°C/min under a flow of helium at 25 ml/min. The obtained TPD profile was recorded with a thermal conductivity detector (TCD) (Micromeritics 2750). For the titration with benzoic acid, 0.025 g of catalyst was dispersed in the mixed solution of DI water and ethanol (4:1 v/v) in the Erlenmeyer flask. The phenolphthalein was added 1-2 drops to the solution as an indicator. Then, 0.5 M of benzoic acid was added dropwise until the pink colour of the solution disappeared. The volume of benzoic acid used was measured to calculate the amount of basic site (mmol/g-catalyst).

3. Results and discussion

3.1. Effect of type of transition metal oxide supported on magnesium oxide catalyst and the reaction temperature

3.1.1. Glucose conversion. The catalytic activity of various transition metal oxides supported on magnesium oxide under mild conditions is shown in figure 1. The reaction was performed at 100°C, 120°C, and 140°C for 1 h under N_2 at 0.4 MPa. It was found that the conversion of all catalysts was above 50% and increased to 100% with an increase in the reaction temperature. This suggested that the increase in reaction temperature enhanced the reaction of glucose conversion [15]. The ZnCTAB/MgO exhibited the highest activity for glucose conversion (64%), while the others follow the trend as NiCTAB/MgO > CoCTAB/MgO > CuCTAB/MgO.

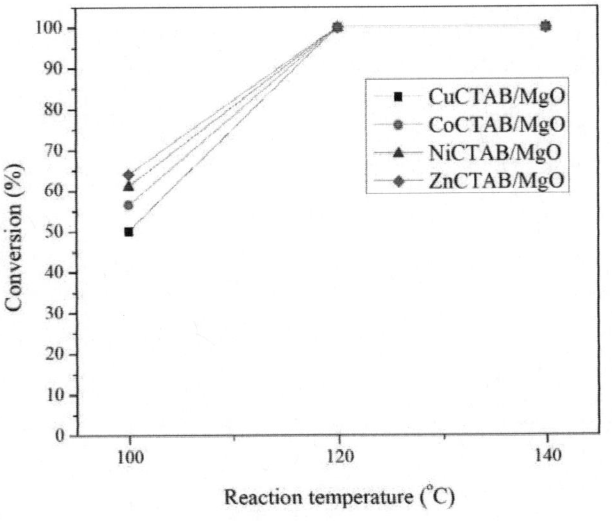

Figure 1. Effect of type of transition metal oxide supported on magnesium oxide catalyst and the reaction temperature on glucose conversion
(Glucose 0.54 g in water 30 ml, catalyst 0.36 g, N_2 0.4 MPa, 1 h, 400 rpm).

3.1.2. Lactic acid yield. Figure 2 describes lactic acid yield from various transition metal oxides supported on magnesium oxide at 100°C, 120°C, and 140°C. An increase in reaction temperature increases the lactic acid yield for all catalysts as has been previously reported [13, 16, 17]. When the reaction was performed at a lower temperature (100°C and 120°C), the NiCTAB/MgO exhibited the highest selectivity in lactic acid production compared with other catalysts to an obtained lactic acid yield of 2.6% and 5%, respectively. In addition, ZnCTAB/MgO gave the highest lactic acid yield (12%) at a high reaction temperature of 140°C.

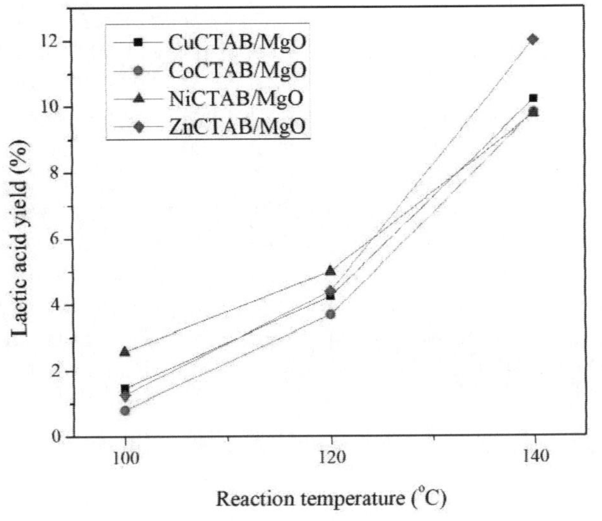

Figure 2. Effect of type of transition metal oxide supported on magnesium oxide catalyst and the reaction temperature on lactic acid yield
(Glucose 0.54 g in water 30 ml, catalyst 0.36 g, N_2 0.4 MPa, 1 h, 400 rpm).

3.2. Catalyst characterization

Catalysts were characterized by XRD to determine the phase of the catalyst. The XRD patterns of various metal oxides supported on the magnesium oxide catalyst are reported in figure 3.

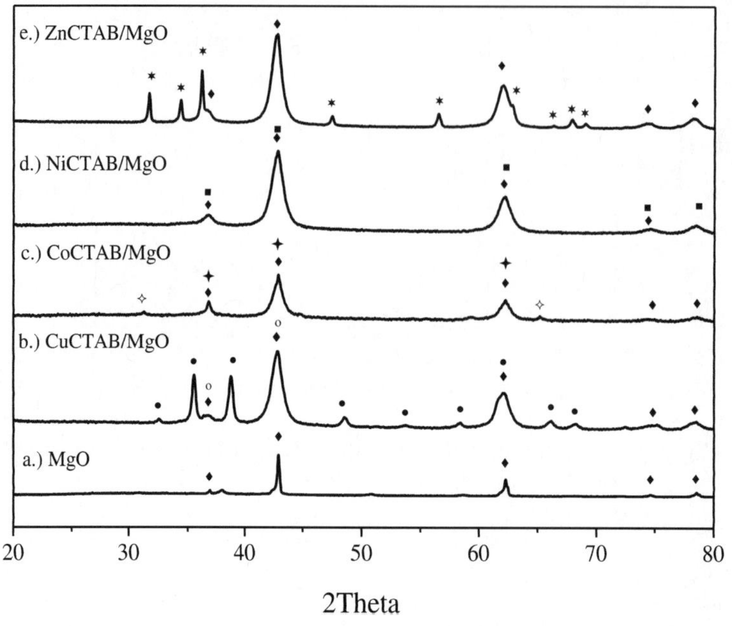

Figure 3. XRD patterns of various divalent metal oxides supported on magnesium oxide catalyst. The different species are shown as MgO (♦), CuO (●), Cu_2O (O), CoO (✦), Co_3O_4 (✧), NiO (■) and ZnO (✱).

All catalysts displayed the characteristic peaks of MgO at 36.9°, 42.9°, 62.3°, 74.7°, and 78.6°. For CuCTAB/MgO catalyst (figure 3b), the oxide species of copper formed showed both the Cu_2O and CuO characteristics. The peaks for CuO were observed at 36.4°and 39.5°. For the CoO peak, (figure 3c) they are 36.8°, 42.8° and 62.2° with minor peaks at 31.3° and 65.2° suggested to be the Co_3O_4, while NiCTAB/MgO (figure 3d) showed all characteristics peaks of NiO which were aligned with the peak of the magnesium oxide at 36.8°, 42.8° and 62.2°. And the characteristic peaks of ZnCTAB/MgO (figure 3e) showed the ZnO characteristics at 31.7°, 34.4°, 36.2°, 36.7°, 42.8°, 47.4°, 56.5°, and 62.0°. From XRD patterns, it is noticeable that the peak of all metal oxide species was a coincidence with the peak of magnesium oxide indicating that the CuO, CoO, NiO, and ZnO were supported successfully on the MgO. The catalyst has a mixed oxide phase with no impurity presented [18].

The basicity of the catalyst was investigated from the temperature programmed desorption of carbon dioxide (CO_2-TPD) and the titration with benzoic acid. The CO_2-TPD profile and the amounts of basicity calculated from CO_2-TPD and benzoic acid titration are shown in figure 4 and table 1, respectively.

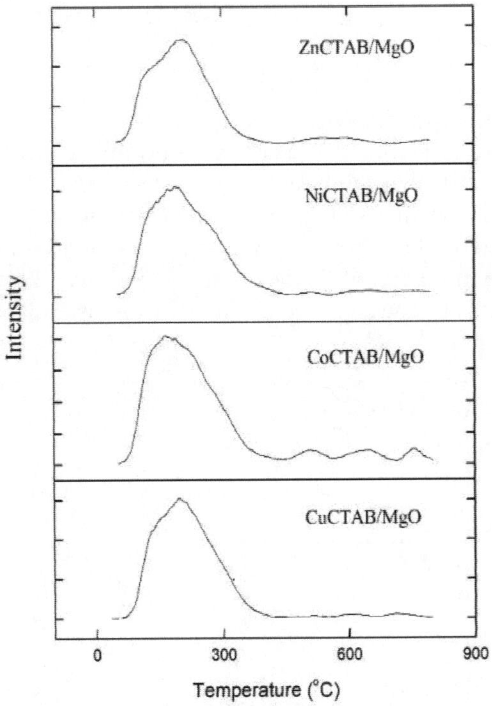

Figure 4. CO_2-TPD profile of various divalent metal oxides supported on magnesium oxide catalyst.

Table 1. Amounts of basicity which was calculated from CO_2-TPD and benzoic acid titration.

Catalyst	Amounts of basicity	
	CO_2-TPD (μmol CO_2/g cat)	Benzoic acid titration (mmol/g cat)
CuCTAB/MgO	462.61	3.8
CoCTAB/MgO	259.67	1.9
NiCTAB/MgO	343.34	2.6
ZnCTAB/MgO	403.17	4.8

From CO_2-TPD profile, the basic strength of all catalysts was similar showing the CO_2 desorption peaks at the temperature lower than 300°C, indicating that metal oxide supported on magnesium oxide catalysts has weak basic strength. The trend in the amounts of basicity calculated from the CO_2-TPD were in the order of CuCTAB/MgO > ZnCTAB/MgO > NiCTAB/MgO > CoCTAB/MgO, while the trend obtained from the benzoic acid titration method is ZnCTAB/MgO > CuCTAB/MgO > NiCTAB/MgO > CoCTAB/MgO. This indicated that the basicity of ZnCTAB/MgO increased when the catalyst was added into the water as confirmed by Sengupta G. (1978) which reported the chemisorption of water on the surface of ZnO at 150°C. They found that at the surface of the catalyst, both Zn^+ and O^- ions act as water adsorption sites. The dissociation of a water molecule into the •H and •OH radicals chemisorbed to ions on the ZnO surface generates the Zn(OH) species on the surface shown below[19].

$$ZnO + H_2O \rightarrow \overset{OH}{\overset{|}{Zn}}\ \overset{H}{\overset{|}{O}} \tag{3}$$

The OH group generated at the surface may increase the amount of basicity, resulting in the highest activity found in the case of ZnCTAB/MgO. Choudhary et al. (2015) reported the mechanism of glucose conversion into lactic acid that the basicity of catalyst played a crucial role in the reaction pathway including the hydrolysis, isomerization, retro-aldol reaction, and the transformation into lactic acid. The reducibility of the catalyst has a role in the formation of pyruvaldehyde-complex form which is the intermediate prior to the formation of the lactic acid [13]. To understand the reducibility of the catalyst, the H_2-TPR was employed. The analyzed profiles of various transition metal oxides supported on magnesium oxide are shown in figure 5.

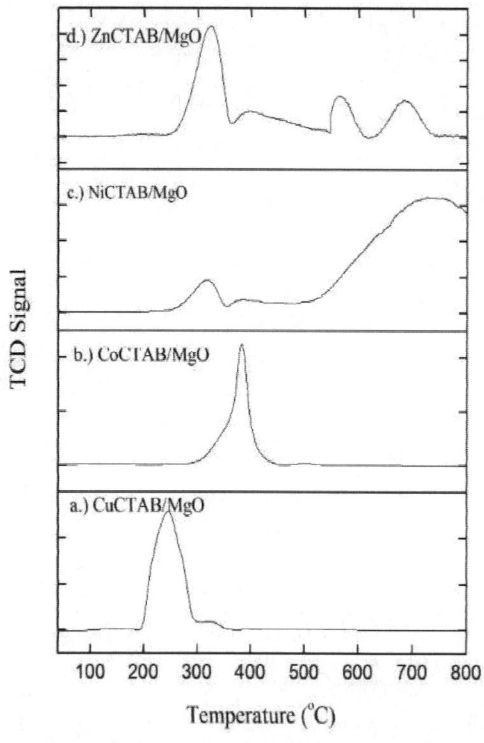

Figure 5. H_2-TPR profiles of various transition metal oxides supported on magnesium oxide catalyst.

The CuCTAB/MgO exhibited a peak at a lower temperature than the other catalysts. The peak at the temperature of 220°C attributes to a well-dispersed CuO phase which can be easily reduced and at the peak of 270°C is reflected the reduction of larger CuO particles. Another small peak at 320°C is due to a low metal dispersion [20]. There is a reduction peak at 360-400°C for CoCTAB/MgO, which is ascribed to the reduction of Co_3O_4 [21]. For NiCTAB/MgO, the small reduction peak occurred at a temperature below 400°C, illustrating the bulk nickel oxide with low interaction with the support. The peak at temperatures higher than 700°C describes the incorporation of NiO with MgO support and the formation of NiO–MgO solid solution [22]. In addition, the ZnCTAB/MgO has a reduction peak at 330°C, 380°C, 560°C, and 680°C, which means that it is hardly reduced compared to the CuCTAB/MgO. However, the glucose conversion and lactic acid yield from ZnCTAB/MgO was higher than the CuCTAB/MgO catalyst. This observation suggests that the basicity of catalyst affects the lactic acid production from glucose more than the reducibility of the catalyst.

4. Conclusions

The investigation on the effect of different divalent transition metal oxide supported on magnesium oxide using CTAB as a capping agent on the glucose conversion to lactic acid found that the ZnCTAB/MgO exhibited the highest activity with a glucose conversion of 64% at 100°C, 0.4 MPa, and 1 h reaction time, whereas the lactic acid yield at 1 h of 12% at 140°C and 0.4 MPa was achieved. This catalytic system consumes less energy, lowers reactor corrosion, and is considered environmentally friendly because of the absence of alkali catalyst typically used in such reaction. A high activity achieved by the ZnCTAB/MgO was suggested to be caused by high basicity produced via water chemisorption on the catalyst's surface.

5. References

[1] Adam YS, Fang Y, Huo Z, Zeng X, Jing Z, Jin F. Production of carboxylic acids from glucose with metal oxides under hydrothermal conditions. *Research on Chemical Intermediates*. 2013;**41**(5):3201-11.

[2] Gao C, Ma C, Xu P. Biotechnological routes based on lactic acid production from biomass. *Biotechnology Advances*. 2011;**29**(6):930-9.

[3] Si W-J, Yang L, Weng Y-X, Zhu J, Zeng J-B. Poly(lactic acid)/biobased polyurethane blends with balanced mechanical strength and toughness. *Polymer Testing*. 2018;**69**:9-15.

[4] Jin F. Application of hydrothermal reactions to biomass conversion. He L-N, Rogers RD, Su D, Tundo P, Zhang ZC, editors. London: Springer-Verlag Berlin Heidelberg; 2014. 409 p.

[5] Li S, Deng W, Li Y, Zhang Q, Wang Y. Catalytic conversion of cellulose-based biomass and glycerol to lactic acid. *Journal of Energy Chemistry*. 2018.

[6] Yan X, Jin F, Tohji K, Moriya T, Enomoto H. Production of lactic acid from glucose by alkaline hydrothermal reaction. *Journal of Materials Science*. 2007;**42**(24):9995-9.

[7] Wang Y, Jin F, Sasaki M, Wahyudiono, Wang F, Jing Z, et al. Selective conversion of glucose into lactic acid and acetic acid with copper oxide under hydrothermal conditions. *AIChE Journal*. 2013;**59**(6):2096-104.

[8] Huo Z, Fang Y, Ren D, Zhang S, Yao G, Zeng X, et al. Selective Conversion of Glucose into Lactic Acid with Transition Metal Ions in Diluted Aqueous NaOH Solution. *ACS Sustainable Chemistry & Engineering*. 2014;**2**(12):2765-71.

[9] Gao X, Zhong H, Yao G, Guo W, Jin F. Hydrothermal conversion of glucose into organic acids with bentonite as a solid-base catalyst. *Catalysis Today*. 2016;**274**:49-54.

[10] Xue Y, Jin F, Yoshikawa K. Hydrothermal Lactic Acid Production from Glucose over Feldspars as Solid Base Catalysts in Water. *Energy Procedia*. 2014;**61**:2474-7.

[11] Duo J, Zhang Z, Yao G, Huo Z, Jin F. Hydrothermal conversion of glucose into lactic acid with sodium silicate as a base catalyst. *Catalysis Today*. 2016;**263**:112-6.

[12] Onda A, Ochi T, Kajiyoshi K, Yanagisawa K. Lactic acid production from glucose over activated hydrotalcites as solid base catalysts in water. *Catalysis Communications*. 2008;**9**(6):1050-3.

[13] Choudhary H, Nishimura S, Ebitani K. Synthesis of high-value organic acids from sugars promoted by hydrothermally loaded Cu oxide species on magnesia. *Applied Catalysis B: Environmental*. 2015;**162**:1-10.

[14] Abd.Rahman N, Hasan M, Hussain M, Jahim J. Determination of Glucose and Fructose from Glucose Isomerization Process by High performance Liquid Chromatography with UV Detection. *Modern applied science*. 2009;**2**.

[15] Kitajima H, Higashino Y, Matsuda S, Zhong H, Watanabe M, Aida TM, et al. Isomerization of glucose at hydrothermal condition with TiO 2 , ZrO 2 , CaO-doped ZrO 2 or TiO 2 -doped ZrO 2. *Catalysis Today*. 2016;**274**:67-72.

[16] Wang Y, Deng W, Wang B, Zhang Q, Wan X, Tang Z, et al. Chemical synthesis of lactic acid from cellulose catalysed by lead(II) ions in water. *Nat Commun*. 2013;**4**:2141.

[17] Onda A, Ochi T, Kajiyoshi K, Yanagisawa K. A new chemical process for catalytic conversion of d-glucose into lactic acid and gluconic acid. *Applied Catalysis A: General*. 2008;**343**(1-2):49-54.
[18] Mastuli MS, Kamarulzaman N, Kasim MF, Sivasangar S, Saiman MI, Taufiq-Yap YH. Catalytic gasification of oil palm frond biomass in supercritical water using MgO supported Ni, Cu and Zn oxides as catalysts for hydrogen production. *International Journal of Hydrogen Energy*. 2017;**42**(16):11215-28.
[19] Sengupta G, Ahluwalia HS, Banerjee S, Sen SP. Chemisorption of water vapor on zinc oxide. *Journal of Colloid and Interface Science*. 1979;**69**(2):217-24.
[20] Zhang Z, Zhou F, Chen K, Fu J, Xiuyang L, Ouyang P. Catalytic In Situ Hydrogenation of Fatty Acids into Fatty Alcohols over Cu-Based Catalysts with Methanol in Hydrothermal Media2017.
[21] Wu M, Fu Y, Zhan W, Guo Y, Guo Y, Wang Y, et al. Catalytic Performance of MgO-Supported Co Catalyst for the Liquid Phase Oxidation of Cyclohexane with Molecular Oxygen. *Catalysts*. 2017;**7**(5):155.
[22] Ranjbar A, Irankhah A, Aghamiri SF. Reverse water gas shift reaction and CO2 mitigation: nanocrystalline MgO as a support for nickel based catalysts. *Journal of Environmental Chemical Engineering*. 2018;**6**(4):4945-52.

Acknowledgment
The authors gratefully thank Global Green Chemicals Public Company Limited (GGC) for financial support.

Reactive Magnetron Sputter Deposition of Copper on TiO_2 Support for Photoreduction of CO_2 to CH_4

Nantiya Jantarasorn[1], Okorn Mekasuwandumrong[2], Peter Kelly[3] and Piyasan Praserthdam[1]

[1]Center of Excellence on Catalysis and Catalytic Reaction, Department of Chemical Engineering, Faculty of Engineering, Chulalongkorn University, Bangkok, 10330, Thailand
[2]Department of Chemical Engineering, Faculty of Engineering and Industrial Technology, Silpakorn University, Nakorn Pathom 73000, Thailand
[3]Surface Engineering Group, School of Engineering, Manchester Metropolitan University, Manchester, M1 5GD, United Kingdom

Corresponding author: piyasan.p@chula.ac.th

Abstract. In this work, nanocrystalline Cu/TiO2 catalysts have been synthesized by using pulsed direct current (DC) reactive magnetron sputtering of Cu targets in an Ar atmosphere onto P25-TiO_2 support. The oscillating bowl was used to make the uniform coating on the substrate. The Cu doping content was varied by adjusting the coating time. The thus-obtained catalysts were characterized by using the X-ray diffraction (XRD), UV-Vis spectroscopy, scanning electron microscopy (SEM), and energy dispersive X-ray spectroscopy (EDX). The photocatalytic activities of all catalysts were studied via the photocatalytic reduction of CO_2 and H_2O to CH_4 under UV irradiation, and compared with the pure TiO_2 support and conventional-impregnation-made Cu/TiO_2. The results showed that the photocatalytic performance of sputtering-made Cu/TiO_2 catalyst was much better than the pure TiO_2 support. Therefore, reactive magnetron sputtering was a promising technique for deposition of metal onto the support and use as the catalytic process.

1. Introduction

Presently, the environmental problem from the emission of pollutant gases became much more serious. CO_2 (Carbon dioxide) is the main product from human activates such as combustion of industrial fuels and vehicle engine. The solution to the problems is reduction of releasing gas into the atmosphere or takes advantage from it. CO_2 can used for generate the important products by direct and indirect process. In direct process, it is widely used in cooler system, extinguisher, dry ice and so on. In indirect process, it is used for generate chemical commercial such as urea, polycarbonate, methanol, acetic acid, formaldehyde and so on [1-2].

Photocatalytic reduction of CO_2 is one of the interesting processes. This technology will convert photocatalytic CO_2 reduction with water as a reducing agent into hydrocarbon fuels and useful chemical compounds such as carbon Monoxide (CO), methane (CH_4), methanol (CH_3OH), formic Acid (HCOOH) and formaldehyde (HCHO) by irradiation with UV light at room temperature and atmospheric pressure [3-4]. Photo-chemical technique is more attractive because of its relatively low cost, environmental friendly, generate none pollution, no requirement of thermal energy. Moreover, it

Content from this work may be used under the terms of the Creative Commons Attribution 3.0 licence. Any further distribution of this work must maintain attribution to the author(s) and the title of the work, journal citation and DOI.
Published under licence by IOP Publishing Ltd

is easy for control process because photochemical reaction need only light to activate substrate to generate products.
Many photocatalytic CO_2 reduction catalysts have been attentively studied such as TiO_2 , ZnO, ZrO_2 , CeO_2 , WO_3, SnO_2, GaP, SiC, $SrTiO_3$ [5]. Among various semiconductor materials, the most extensively used photocatalysts is TiO_2, because its high stability, high photosensitivity, non-toxic nature, wide availability and low cost [6]. However, TiO_2 has wider bandgap, high electron-hole recombination rate and weak CO_2 adsorption. Therefore, it has very low photocatalytic activity for CO_2 reduction in the water. So, it is interesting to modify the crystal structure, particle size, electronic structure, lifetime of charge carrier, electron-hole (e-h) recombination probability of TiO_2 for increase photocatalytic activity for CO_2 conversion [8]. There are several pathway for improve its quality such as doping with metals or non-metals. Some metal such as platinum (Pt), silver (Ag), copper (Cu), nickel (Ni) or metal oxide such as copper oxide (Cu_XO), nickel oxide (NiO) were deposited on semiconductor. For increased lifetime of the photogenerated electrons and hole via effective charge carrier separation and retardation of electron-hole recombination rate and increasing quantum yield [7-8]. In several studies, Cu nanoparticles deposited on the TiO_2 surface can reduced the band gap energy of TiO_2 and shifted the band edge from the UV region to the visible region (strongly absorb visible light) and facilitate the separation of photoexcited e^- h^+ pairs that showed effectiveness in CO_2 photoreduction [9-10].
Generally, deposition of metal on TiO_2 surface can be mostly done by conventional impregnation method, which required the solution of metal precursor, heat treatment and generate waste. Reactive magnetron sputtering offers the new direct metal deposition method on catalyst powder, which can be done in one-step and no waste generate in the process [11].
In this research, Cu/TiO_2 catalysts were prepared by using reactive magnetron sputtering to deposit the Cu on TiO_2 surface. The obtained catalysts were characterized by using the X-ray diffraction (XRD), UV-Vis spectroscopy, scanning electron microscopy (SEM), and energy dispersive X-ray spectroscopy (EDX). The photocatalytic activities of all catalysts were studied via the photocatalytic reduction of CO_2 and H_2O to CH_4 under UV irradiation and compare it with the catalyst prepared by conventional impregnation method.

2. Experiment

2.1. Materials
Commercial Degussa P25 nanocatalyst from AERO-XIDE1 was sourced as a TiO_2 powder. Copper(II) nitrate trihydrate ($Cu(NO_3)_2 \cdot 3H_2O$,>98% purity from Aldrich), and Cu target (Copper Metal, 99.5% purity) were used as a chemicals in the surface modification process.Carbon dioxide (CO_2 ,≥99.999 %) was used as an oxidizing agent in photocatalytic reaction.

2.2. Surface modification of TiO_2
There are two preparation methods for Cu/TiO_2 catalysts including incipient wetness impregnation method and reactive magnetron sputtering method. In incipient wetness impregnation methods, 2 g of TiO_2 was doped with an aqueous solution of copper (II) nitrate trihydrate ($Cu(NO_3)_2 \cdot 3H_2O$) via different the weight ratios of dopant/ TiO_2 at 1 (Im1) and 3 (Im2) wt%, respectively. The solution of the metal precursor was slowly dropped onto the TiO_2 support to obtain the desired ratio and dried in oven 110 °C overnight before calcined in a box furnace at 400 °C, heating rate 10 °C/min for 2 h with air flow.
In reactive magnetron sputtering method, 2 g of TiO_2 was loaded in sputtering chamber. Synthesis condition including power 200W, voltage 350V and frequency 200kHz were used. Reactive sputtering of Cu target was done in argon atmosphere onto P25-TiO_2 support. The oscillating bowl was used to make the uniform coating on the substrate. The sputtering time was varied at 2.5 (Sp1) and 7.5 (SP2) min respectively.

2.3. Photocatalytic reaction

The photocatalytic reduction of CO_2 with water were carried out in a photoreactor system. UV-light bulbs (Philips' Germi-cidal Ozone UV Quartz Glass UVC Bulb : 16 watt, 6 bulbs) were install around a cylindrical quartz reactor . A certain amount of each catalyst (0.5 g) dispersed into a stirred slurry reactor (SSR), which contained 150 mL of deionized water. After the reaction system was already, compressed CO_2 (≥99.999 %) was presented into the system from inlet tube in the cylindrical quartz reactor. By using a mass flow controller at a flow rate of 100 ml min^{-1} for 30 min. to purge air and saturate the solution. The reactor is closed during the reaction, and a magnetic stirrer agitated the catalyst-suspended solution throughout the experiment. The photocatalytic reaction was started by turned on the UV light, and irradiation was continued for 6 h.
The resultant gas samples were analyzed online by using, a GC-14B (Shimadzu) gas chromatograph equipped with a flame ionization detector (FID) and a Porapak-Q column for hydrocarbon analysis.

3. Results and discussion

3.1. Physiochemical properties of Cu/TiO$_2$ catalysts

Table 1. The physical properties of thus-obtained catalysts.

Sample	Wt% of Cu		Crystallite size (nm)			%Rutile
	ICP	EDX	Anatase	Rutile	CuO	
P25	-	-	20	23	-	21
Im1	0.68	3.42	18	22	-	20
Im2	2.47	5.67	17	21	-	18
Sp1	0.73	4.89	19	25	n.d.	22
Sp2	2.21	6.15	19	23	45.00	25

In this paper, nanocrystalline Cu/TiO_2 catalysts have successfully been prepared by using magnetron sputtering method. The physical properties of thus-obtained catalysts are summarized in table 1. The Cu doping content was measured by ICP and EDX techniques. Increasing of sputtering time from 2.5 to 7.5 min resulted in the increase of Cu doping content from 0.73 to 2.21 wt%. It can be seen that the Cu loading content increased linearly with increasing of the sputtering time, which suggested that the Cu sputtering from Cu target was directly deposited on the TiO_2 support. To compare the sputtering-made catalyst, two catalysts with similar Cu content loading were prepared by using conventional impregnation method.
Figure 1 shows the XRD pattern of P25-TiO_2 support and Cu/TiO_2 catalysts prepared by magnetron sputtering and conventional impregnation methods. All catalysts exhibited the main characteristic peak of anatase TiO_2 phase ($2\theta = 25.34°$) with small additional peak of rutile TiO_2 phase ($2\theta = 27.42°$) [12]. Small peak of CuO was also observed in sputtering-made samples, which due to the formation of larger CuO particles comparing with impregnation one. The crystallite size and phase contents of the all samples were calculated by their XRD patterns according to the methods of the Debye–Scherrer equation and Spurr (as summarized in table 1). It can be seen that the TiO_2 crystallite size and rutile phase content did not change much after addition of Cu by both sputtering and conventional impregnation methods.

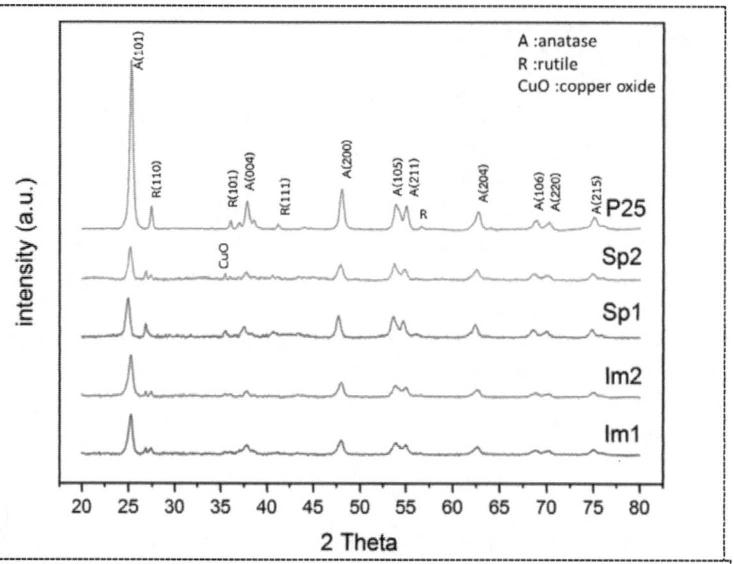

Figure 1. XRD pattern of P25-TiO_2 support and Cu/TiO_2 prepared by magnetron sputtering and conventional impregnation methods.

Figure 2 shows UV-Vis spectra of all catalysts. Addition with Cu with both methods resulted in an increase in the absorption of the TiO_2 photocatalyst in the visible region (400-700nm). Moreover, increasing of Cu doping content also improved the visible light adsorption [12-13]. The lowering of band gap by the doping may eventually leads to Fermi level reducing, improving the photosensitivity of the catalysts. The energy band gaps (E_{bg}) of all the catalyst samples were obtained from the extrapolation of Tauc plot to the abscissa of photon energy and results are presented in table 2. The band gap energy of P-25 (3.18 eV) decreased to 2.87 and 3.01 eV after doping with Cu by impregnation and sputtering methods, respectively.

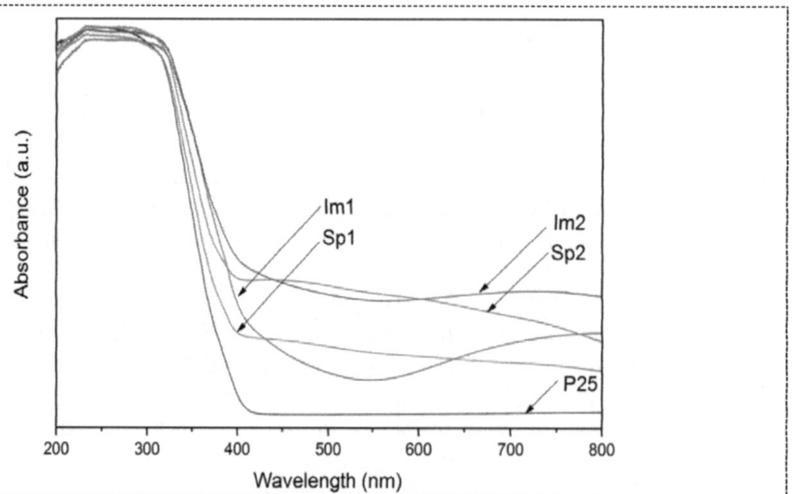

Figure 2. Presents the UV–Vis absorbance spectra of the P25-TiO_2 and Cu/TiO_2 prepared by using the magnetron sputtering and conventional impregnation methods.

SEM was employed to investigate the morphology of the as-synthesized Cu/TiO_2 catalysts as presented in figure 3. The micrographs clearly showed the agglomeration of nanosized spherical particles. There are not much difference between the Cu/TiO_2 catalysts prepared by both methods. To investigate the elemental composition of thus-obtained catalysts, EDX was employed as also summarized in table 1. The similar trend with ICP was also observed but the Cu content determined by EDX was much higher than those determined by ICP methods. This probably due to the formation of CuO by both methods mainly occurred on the TiO_2 surface and not deposited deep through the TiO_2 pore.

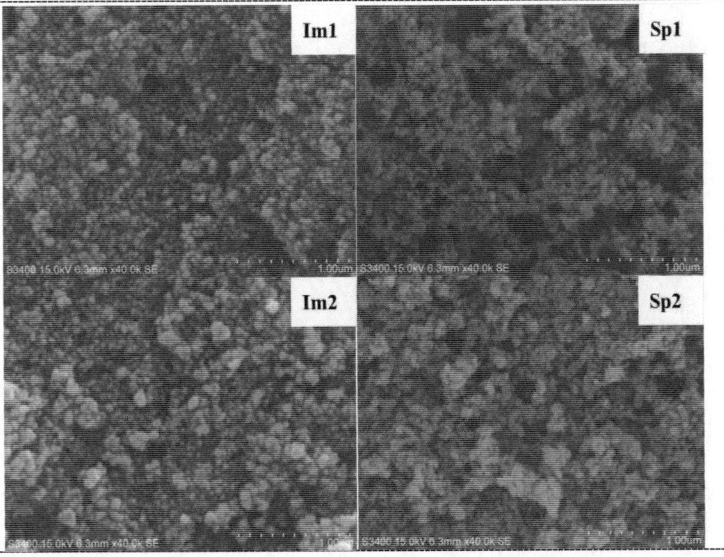

Figure 3. SEM images of all Cu/TiO_2 catalysts.

3.2. Photocatalytic activity

Table 2. The band gap energy of thus-obtained catalysts and CH_4 production rate.

Sample	Band gap energy (eV)	CH_4 production rate (μmole.gcat^{-1}.h^{-1})
P25	3.18	0.27
Im1	2.81	0.74
Im2	2.87	0.88
Sp1	3.09	1.25
Sp2	3.01	1.86

The catalytic activity of Cu/TiO_2 catalysts prepared by magnetron sputtering and conventional impregnation methods were tested in photocatalytic reduction of CO_2 and H_2O to methane at atmospheric temperature and pressure for 6 hours. The CH_4 production rates are summarized in table 2. Lower CH_4 production rates of P25 were due to the fast recombination of photo-generated electron-hole pairs. Doping with Cu into TiO_2 resulted in the improvement of CH_4 production rate [13-14]. In photocatalytic process, electron (e^-) and hole (h^+) are generated during the photocatalytic process. Due to the short lifetime, they can recombine immediately or can participate in oxidation and reduction process. The holes can oxidize H_2O or H_2 while the photo-excited electrons can reduce CO_2 to form CO, CH_4 and CH_3OH. When doping with Cu metals on the TiO_2 surface, the Cu can trap the CB electrons, which prolong the electron-hole recombination process and improve the photocatalytic activity[13-14].

4. Conclusion

Reactive magnetron sputtering technique has successfully been used to deposit nanocrystalline CuO on the TiO_2 support. The Cu doping content increased linearly with prolong sputtering time. Doping with Cu did not changed the catalyst morphology and TiO_2 crystallite size, while, it increased the adsorption at visible light range[12-13]. The photocatalytic reduction performance of sputtering-made Cu/TiO_2 catalyst was much better than the pure TiO_2 support. Therefore, reactive magnetron sputtering was a promising technique for deposition of metal onto the support and use as the catalytic process.

5. References

[1] M. Takeuchi, Y. Sakamoto, and S. Niwa 2001 Study on CO2 global recycling system *Science of the total environment* vol. 277 no. 1-3 pp. 15-19

[2] I.-H. Tseng, W.-C. Chang, and J. C. Wu 2002 Photoreduction of CO2 using sol–gel derived titania and titania-supported copper catalysts *Applied Catalysis B: Environmental* vol. 37 no. 1 pp. 37-48

[3] M. Cokoja, C. Bruckmeier, B. Rieger, W. A. Herrmann, and F. E. Kuehn 2011 Transformation of carbon dioxide with homogeneous transition-metal catalysts: a molecular solution to a global challenge? *Angewandte Chemie International Edition* vol. 50 no. 37 pp. 8510-8537

[4] W. Qingli *et al.* 2015 Photoreduction of CO2 using black TiO2 films under solar light *Journal of CO2 Utilization* vol. 12 pp. 7-11

[5] K. Wenderich and G. Mul 2016 Methods, mechanism, and applications of photodeposition in photocatalysis: a review *Chemical reviews* vol. 116 no. 23 pp. 14587-14619

[6] H. Yamashita, A. Shiga, S.-i. Kawasaki, Y. Ichihashi, S. Ehara, and M. Anpo 1995 Photocatalytic synthesis of CH4 and CH3OH from CO2 and H2O on highly dispersed active titanium oxide catalysts *Energy conversion and management* vol. 36 no. 6-9 pp. 617-620

[7] S. Das and W. W. Daud 2014 A review on advances in photocatalysts towards CO2 conversion *Rsc Advances* vol. 4 no. 40 pp. 20856-20893

[8] Y. Ohama and D. Van Gemert 2011 *Application of titanium dioxide photocatalysis to construction materials: state-of-the-art report of the RILEM Technical Committee* 194-*TDP* Springer Science & Business Media

[9] H. Yan, T. Zhao, X. Li, and C. Hun 2015 Synthesis of Cu-doped nano-TiO2 by detonation method *Ceramics International* vol. 41 no. 10 pp. 14204-14211

[10] H. Li, C. Li, L. Han, C. Li, and S. Zhang 2016 Photocatalytic reduction of CO2 with H2O on CuO/TiO2 catalysts *Energy Sources, Part A: Recovery, Utilization, and Environmental Effects* vol. 38 no. 3 pp. 420-426

[11] M. Ratova, R. Klaysri, P. Praserthdam, and P. J. Kelly 2018 Visible light active photocatalytic C-doped titanium dioxide films deposited via reactive pulsed DC magnetron co-sputtering: Properties and photocatalytic activity *Vacuum* vol. 149 pp. 214-224

[12] H. W. Nasution, E. Purnama, S. Kosela, and J. Gunlazuardi 2005 Photocatalytic reduction of CO2 on copper-doped Titania catalysts prepared by improved-impregnation method *Catalysis Communications* vol. 6 no. 5 pp. 313-319

[13] Y. Li, W.-N. Wang, Z. Zhan, M.-H. Woo, C.-Y. Wu, and P. Biswas 2010 Photocatalytic reduction of CO2 with H2O on mesoporous silica supported Cu/TiO2 catalysts *Applied Catalysis B: Environmental* vol. 100 no. 1-2 pp. 386-392

[14] B. Xin, P. Wang, D. Ding, J. Liu, Z. Ren, and H. Fu 2008 Effect of surface species on Cu-TiO2 photocatalytic activity *Applied Surface Science* vol. 254 no. 9 pp. 2569-2574

Acknowledgments

This research is supported by the Ratchadapisek Sompoch Endowment Fund (2016), Chulalongkorn University (CU-59-006-IC) and Newton Mobility Grants.

The Recovery Process of Used Hydrocarbon Solvent by Vacuum Distillation

Apiched Savasdipol[1], Nophawan Paradee[2] and Sumonman Niamlang[1*]

[1] Department of Material and Metallurgical Engineering, Faculty of engineering, Rajamangala University of Technology Thanyaburi, Pathumtani, Thailand
[2] Department of Chemistry, Faculty of Science, King Mongkut's University of Technology Thonburi, Bangkok, Thailand

*Corresponding author: sumonman.n@en.rmutt.ac.th

Abstract. In polymer production industry, a large amount of hydrocarbon solvent (HC solvent) are used to clean the reactor. Normally, used HC solvent are managed by Incineration process at high temperature (around 700 – 1200°C) which cause a large amount of waste gas explosion such as CO2. To manage the used solvent for energy saving and global warming reason, used HC solvent should be recycled by refining process. The aliphatic hydrocarbon, EXXSOL D80, was selected as the model HC solvent in this research work. The vacuum distillation was selected as the suitable refining process. To determine the optimum refining condition, the effect of vacuum distillation pressure (13, 15, 20, 25 and 30 kPa) on % yield, purity and assessment of economic value was studied. The purity of used solvent and %carbon atom content were confirmed by gas chromatography-mass spectrometer (GC-MS). %Purity of carbon-12, 13 and 14 atom was decreased with increasing distillation pressure. The economic cost was evaluated. At The suitable condition, 170°C 13 kPa, the highest yield of distilled solvent was observed at 77.18%. Thus the recovery used solvent can be reduce cost for purchase and disposal around 33,651.27 USD/year. Finally, vacuum distillation is suitable for recovery used HC solvent.

1. Introduction

The growth of polymer production business is rapidly expanding because polymer products were consumed in daily life. A lot of chemical substances were used as curing agent, color, catalyst, retardant, solvent. Cleaning equipment process is very important process in polymerization production. HC solvent was frequently used in cleaning process for mixer and container.

Aliphatic HC solvent is an organic compound. HC solvents consists of C_{12} – C_{14} saturated hydrocarbons. The physical properties, boiling point, specific gravity (at 15.6 °C) and vapour pressure (at 20 °C) of HC solvent is 209 – 228 °C, 0.80 and 0.2 mmHg. The flash point is 76.7 °C and auto-ignition temperature is 216 °C [1].

In this research, the recovery process of used HC solvent was studied by distillation process. Vacuum distillation was interested for recovery used HC solvent because vacuum distillation can reduce pressure to very low, thus the rate of distillation is increased [2] and the equipment can be easily set up. The influence of distillation pressure on recovery used HC solvent for purity and efficiency was investigated. The chemical and economical evaluation was also interested. The appropriate condition of vacuum distillation will be discussed for the new selection of used HC solvent recovery process.

Content from this work may be used under the terms of the Creative Commons Attribution 3.0 licence. Any further distribution of this work must maintain attribution to the author(s) and the title of the work, journal citation and DOI.
Published under licence by IOP Publishing Ltd

IOP Conf. Series: Materials Science and Engineering **559** (2019) 012018 doi:10.1088/1757-899X/559/1/012018

2. Methodology

2.1 Materials

Reference HC solvent, EXXSOL D80, Dearomatized Fluids, (Union Petrochemical Public Company Limited) were used. Used HC solvent were as received from silicone rubber production cleaning process. The residue silicone oligomer and impurities from used HC solvent were firstly removed by filter (Nylon Monofilament Mesh (NMO) sized of 80 μm).

2.2 Vacuum distillation process

The vacuum distillation conditions were corresponded to Cox's chart vapour pressure [3]. N-dodecane ($C_{12}H_{26}$) was selected as reference solvent for the identification standard parameter in these experiment. The distillation temperature and pressure was 170°C and 13 kPa, respectively. To investigate the effect of distillation pressure, various distillation pressure (13, 15, 20, 25 and 30 kPa) were used. The schematic diagram of the recovery process used HC solvent was shown in figure. 1. The experimental set up was shown in figure. 2

The used HC solvent of 250 ml was poured into round bottom flask (size of 500 mL). The used HC solvent were heated at temperature of 170°C and at distillation pressure of 13, 15, 20, 25 and 30 kPa. The vacuum distillation conditions were tabulated in table 1.

Table 1. The vacuum distillation conditions.

Conditions	Details	Conditions	Details
170°C, 13 kPa	1) Step 1: 30°C, 70 kPa, at start. 2) Step 2: 60°C, 50 kPa, at 20 min. 3) Step 3: 90°C, 30 kPa, at 48 min. 4) Step 4: 110°C, 15 kPa, at 67 min. 5) Step 5: 130°C, 13 kPa, at 85 min. 6) Step 6: 170°C, 13 kPa, for 5 hours.	170°C, 15 kPa	1) Step 1: 30°C, 70 kPa, at start. 2) Step 2: 60°C, 50 kPa, at 20 min. 3) Step 3: 90°C, 30 kPa, at 49 min. 4) Step 4: 110°C, 15 kPa, at 79 min. 5) Step 5: 170°C, 15 kPa, for 5 hours.
170°C, 20 kPa	1) Step 1: 30°C, 70 kPa, at start. 2) Step 2: 60°C, 50 kPa, at 20 min. 3) Step 3: 90°C, 30 kPa, at 48 min. 4) Step 4: 110°C, 20 kPa, at 66 min. 5) Step 5: 170°C, 20 kPa, for 5 hours.	170°C, 25 kPa	1) Step 1: 30°C, 70 kPa, at start. 2) Step 2: 60°C, 50 kPa, at 20 min. 3) Step 3: 90°C, 30 kPa, at 50 min. 4) Step 4: 110°C, 25 kPa, at 70 min. 5) Step 5: 170°C, 25 kPa, for 5 hours.
170°C, 30 kPa	1) Step 1: 30°C, 70 kPa, at start. 2) Step 2: 60°C, 50 kPa, at 18 min. 3) Step 3: 90°C, 30 kPa, at 36 min. 4) Step 4: 170°C, 30 kPa, for 5 hours.		

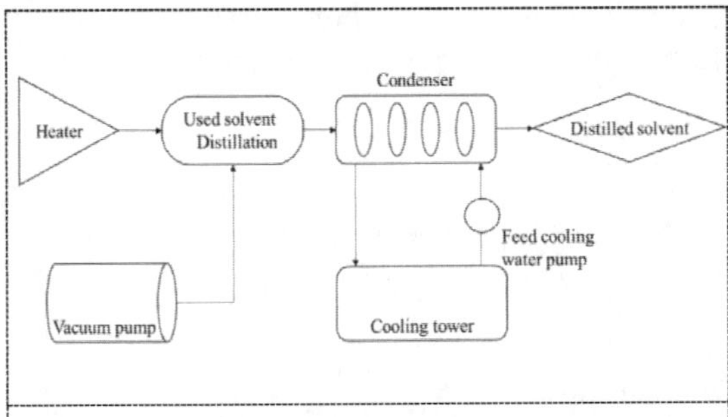

Figure 1. Schematic diagram of recovery used solvent process.

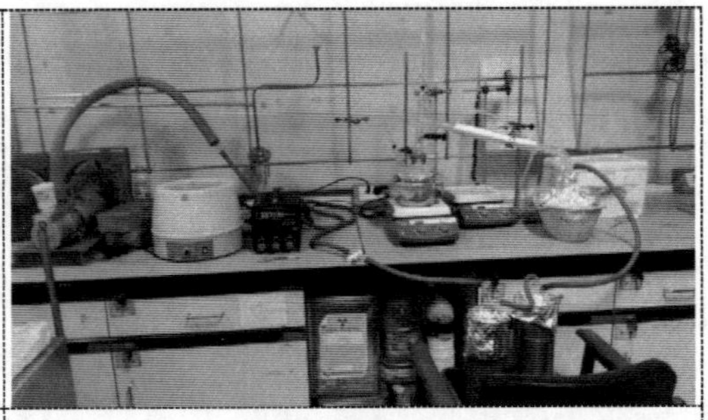

Figure 2. The experimental of vacuum distillation.

2.3 Characterization

To confirm the successful of recovery used solvent process, the purity of distilled solvent was determined by Gas chromatography-mass spectrometer (GC-MS, 7890B-GC & 5977A series GC/MSD, Agilent). Distilled solvent of 0.05 %v/v was used in GC-MS test. The condition of GC-MS is shown in table 2. The functional groups of distilled solvent was characterized by Fourier Transform Infrared Spectroscopy (FT-IR spectroscopy, Perkin Elmer).

Table 2. The condition of GC-MS characterization.

Condition	Parameter	Condition	Parameter
Type of column	HP-5 MS Ultra Inert	Carrier gas	Helium (He)
Column Oven Temp.	75.0 °C	Pressure	100.2 kPa
Hold time (1)	0.5 min	Total flow	161.5 mL/min
Injection temp	250.0 °C	Column flow	1.5 mL/min
Transfer line temp	280.0 °C	Purge flow	10.0 mL/min
Injection volume	1 mL	Split ratio	100.0
Injection mode	Split		

3. Result and discussion

3.1 Effect of pressure on %yield recovery used HC solvent process

The effect of temperature on %yield of recovery used HC solvent was studied from our previous work [4]. The experiments were conducted at pressure of 13 kPa and temperature of 140, 150, 160 and 170°C. The % yield increased with increasing vacuum distillation temperature due to increasing energy for breaking the bond. The maximum %yield was obtained from vacuum distillation temperature of 170°C. At higher than 170°C, the bumping was observed. When the vacuum distillation temperature increased, the %yield of recovery used HC solvent increased corresponding to increasing of energy consumption too [4]. Thus turning the economic worthiness condition is required. To investigate the suitable condition for used solvent waste recovery process, the vacuum distillation process was conducted at 170°C and pressure was varied at 13, 15, 20, 25 and 30 kPa. The %yield of recovery used HC solvent was determined and shown in figure. 3. The % yield decrease with increasing pressure which corresponding to relation between pressure and boiling point of substance. When pressure reduced to very low pressure as the result of the low vapour density was occurred, thus the superficial vapour velocity increased. Due to the effect of pressure on %yield, it is possible to reduce time for distillation [5].

3.2 Distilled solvent characterization

To confirm the functional group of distilled solvent from recovery of used HC solvent process by vacuum distillation. The functional group of distilled HC solvent and reference solvent were first investigated by FT-IR spectroscopy (Figure. 4). The distilled HC solvent shows the similar peak as reference. Thus the distilled HC solvent was successfully recovered by vacuum distillation process.

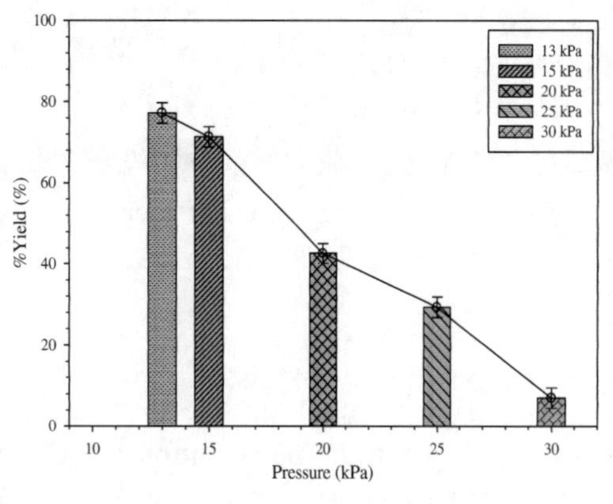

Figure 3. %Yield of distilled solvent at various pressures in vacuum distillation.

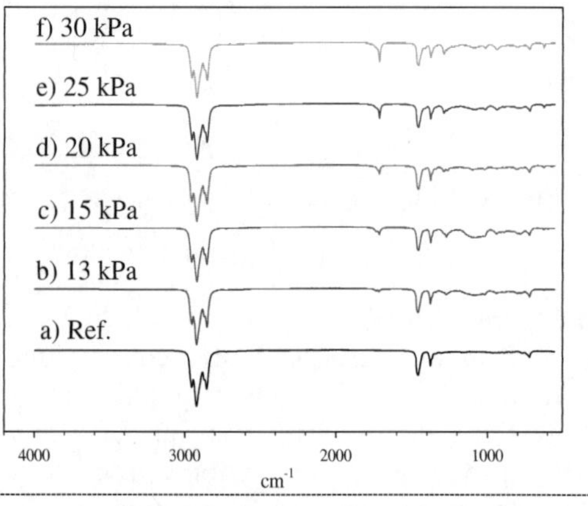

Figure 4. The functional group of a) reference solvent and distilled solvent b) at 13 kPa, c) at 15 kPa, d) at 20 kPa, e) at 25 kPa and f) at 30 kPa.

The important functional group of aliphatic hydrocarbon 2954.98, 2921.89 and 2853.88 cm^{-1} for C-H stretching of alkane, 1457.78 cm^{-1} for C-C stretching of aromatics, 1377.87 cm^{-1} for C-H bending of alkane and 723.09 cm^{-1} for C-H bending of alkene, respectively [6]. From FT-IR spectrum, the peak at 1710.53 cm^{-1} for C=O stretching of carbonyl was observed which corresponding to silicone elastomer production about the one component silicone rubbers, acetoxy silicone. Their designations refer to the by-products generated during the curing process [7]. The structure acetic acid is –COOH and this structure has carbonyl group. High pressure as the result of relative volatility and flow rate of distilled solvent between the components will become smaller compared with low pressures [8]. The GC-MS results might be clarified-these evidences.

3.3 Purity of distilled solvent

To evaluate about %carbon atom content and purity of distilled solvent, the distilled solvent was characterized by GC-MS. Total Ion Chromatogram (TIC) of distilled solvent was investigated and shown in figure. 5 and %carbon atom content of C_{11}, C_{12}, C_{13} and C_{14} is shown in table 3.
From table.3 the %content of C_{12}, C_{13} and C_{14} in distilled HC solvent were characterized by GC-MS. The %carbon atom content of C_{12}, C_{13} and C_{14} under 170°C 13 kPa condition were found highest carbon content when compared with distilled HC solvent at 15, 20, 25 and 30 kPa which corresponding to substance of low molecular weight can be satisfactorily distilled under low pressure provided its high concentration low molecular weight hydrocarbon content [9]. The %purity of distilled HC solvent is shown in table 4.

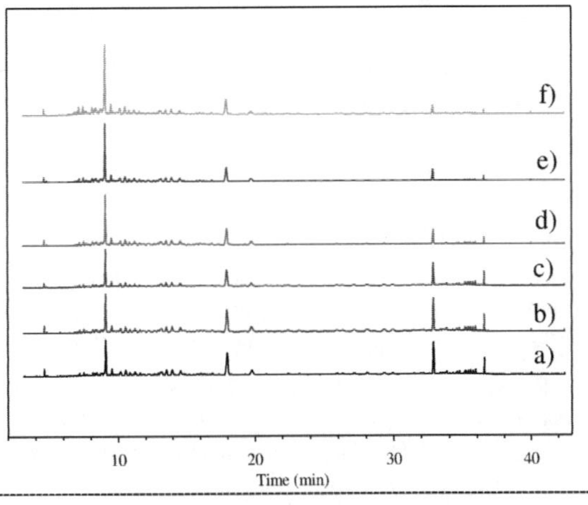

Figure 5. Chromatogram of a) reference solvent and distilled solvent b) at 13 kPa, c) at 15 kPa, d) at 20 kPa, e) at 25 kPa and f) at 30 kPa.

Table 3. The %carbon content of C_{12}, C_{13} and C_{14} in distilled HC solvent.

	%Carbon atom content (%)		
Solvent	**C12**	**C13**	**C14**
EXXSOL D80	35.46	30.52	10.86
Distillation pressure 13 kPa	34.93	27.06	6.02
Distillation pressure 15 kPa	34.18	25.88	5.21
Distillation pressure 20 kPa	34.66	16.71	3.23
Distillation pressure 25 kPa	30.84	16.92	3.24
Distillation pressure 30 kPa	25.03	8.28	1.59

From table. 4 the purity of C_{12}, C_{13} and C_{14} in distilled HC solvent were characterized and calculated. The % purity of distilled HC solvents were compared with reference solvent. From the result at 170°C 13 kPa condition was found C_{12}, C_{13} and C_{14} higher than 15, 20, 25 and 30 kPa, respectively. Because high molecular weight as C_{13} and C_{14} was produced from liquid phase to vapour phase less than low molecular weight as C_{12} due to high pressure was affected to vapour density. As the result of %purity of C_{12} at 13, 15, 20, 25 and 30 kPa were increased with increasing distillation pressure but in C_{13} and C_{14} were decreased with increasing pressure [10].

Table 4. The %purity of distilled HC solvent.

	%Purity (%)		
Condition	**C12**	**C13**	**C14**
Distillation pressure 13 kPa	98.51	88.66	55.43
Distillation pressure 15 kPa	96.39	84.80	47.97
Distillation pressure 20 kPa	97.74	54.75	29.74
Distillation pressure 25 kPa	86.97	55.44	29.83
Distillation pressure 30 kPa	70.59	27.13	14.64

3.4 Economic evaluation of distilled solvent

To study the economic evaluation of distilled HC solvent, the working day is 320 day/year. 200 L of HC solvent was used for cleaning process per day. The cost for purchased HC solvent is 120.76 USD/200L and cost for disposal used HC solvent per year is 4,954.24 USD. Total of cost for purchase and treatment is 43,597.31 USD/year.

The condition 170°C and 13 kPa in recovery process was evaluated. 77.18% of yield was obtained to highest yield of distilled HC solvent. From the result of chemical properties and %carbon atom content is similar with reference HC solvent cause of this condition is suitable for refining process.
Total of recovery solvent per year can be distilled 49,395 L/year. Cost reduction for disposal per year can be reduce to 3,823.65 USD/year. Cost reduction for purchase new HC solvent can be decrease around 29,827.62 USD/year.

4. Conclusion

The suitable condition of recovery used HC solvent by vacuum distillation was studied. The %yield of distilled HC solvents were found in 170°C 13 kPa condition was produced highest yield at 77.18% of recovery process. The chemical composition of distilled solvents were characterized by FT-IR. The distilled solvent shows the similar peak as reference. But at 1710.53 cm-1 was found carbonyl composition when distillation pressure was increased at 15, 20, 25 and 30 kPa which corresponding to silicone elastomer production about the one component silicone rubbers, acetoxy silicone because the curing agent was used acetic acid (-COOH). The %carbon atom content and purity of distilled solvent were determined by GC-MS. The carbon content 34.93% and purity 98.51% of C12 at 170°C 13 kPa was obtained higher than 15, 20, 25 and 30 kPa. But the carbon content and purity of C_{13} and C_{14} were reduced with increasing distillation pressure due to distillation pressure was affected to vapour density. When distillation was used low pressure as the result of low vapour density substance was occurred and effected to high molecular weight (high vapour density) was not occurred. Thus the recovery used HC solvent process will be reduce amount of used solvent for disposal around 49,000 L/year. The cost reduction was evaluated and the cost can be reduced 29,827.62 USD/year. Finally, vacuum distillation is suitable for recovery used HC solvent.

5. References

[1] Material Safety Data Sheet, “EXXSOL D80,” EXXON Company, U.S.A, pp. 1-6, March 1999.
[2] FA Howard and NE Loomis, “Adventages of vacuum distillation,” Oil and Gas Journal, vol. 100(35), pp. 132-133, Jan. 2002.
[3] Kunchana Bunyakiat, Basic calculations in chemical engineering, 8 th ed. Bangkhunthain, Bangkok, 2014, pp. 89-95.
[4] Apiched Savasdipol, Kiyoaki Ishimoto, Natacha Phetyim, Weraporn Pivsa-Art, Sommai Pivsa-Art and Sumonman Niamlang, “Effect of Temperature on Waste Hydrocarbon Solvent Recovery Process,” presented at the STEMa 2018 International conference, Pattaya, Thailand, July 18-20, 2018.
[5] Gary R. Martin and Jason M. Nigg, “Vacuum Unit Pressure Control: Impact on Refinery Profitability,” Petroleum Technology Quarterly,. vol. 6, Summer 2001.
[6] Baghali G. Mathapa and Vesselin N. Paunov, “Self-assembly of cyclodextrin-oil inclusion complexes at the oil-water interface: a route to surfactant-free emulsions” Journal of Materials Chemistry A,1, pp. 10836-10846, 2013.
[7] Silicones, 3rd ed., Handbook of Thermoset Plastics., Günter Lorenz and Andreas Kandelbauer, 2014, pp. 555-575.
[8] Zhi-Yong Liu and Megan Jobson, The Effect of Operating Pressure on Distillation Column Throughput, Computers and Chemical Engineering Supplement, 1999, pp. S831-S834.
[9] Hyungseok Nam, Julius Choi and Sergio C. Capareda, “Comparative study of vacuum and fractional distillation using pyrolytic microalgae (Nannochloropsis oculata) bio oil”, Algal Research 17, 2016, pp. 87-96.
[10] Anna Iakolieva, Segrii Boichenko, Kazimierz Lejda, Oksana Vovk and Iryna Shkilniuk, “Vacuum Distillation of Rapeseed Oil Esters for Production of Jet Fuel Bio-Additives”, Procedia Engineering 187, 2017, pp. 363-370.

Acknowledgments
The authors would like to thanks the financial support from Smart (2015) Services limited and National Science Technology and Innovation Policy Office for support this research.

Numerical Modeling of Industrial-Scale Pulsed Bed Adsorber for Colorant Removal in Sugar Refining

Terasut Sookkumnerd[1]

[1] School of Chemical Engineering, Suranaree University of Technology, Nakhon Ratchasima 30000, THAILAND

Corresponding author: terasut@gmail.com

Abstract. Because pulsed bed adsorbers are preferred when the high fluid flow rate is expected, pulsed bed adsorbers are used in many industries; for example, waste water treatment and food industries. However, few attempts have been focused on modeling of pulsed bed adsorbers. In this paper, the mathematical model of an industrial-scale pulsed bed adsorber has been developed. By using the data on adsorption equilibrium constant, mass transfer and axial dispersion in a lab-scaled fixed-bed adsorber, the numerical analysis of an industrial-scale pulsed bed adsorber for colorant removal in sugar refining process is performed and the numerical results are analyzed to gain the better understanding of this unit operation. The optimum cycle time depended on the fraction of adsorbent being removed and replaced in each pulse; the optimum cycle time increases from 4 hours at 5 % pulse to 12 hours at 14% pulse. In addition, the fluctuation of feed flow rate would affect the performance of pulsed bed adsorbers; for example, the 10% increase of flow rate would make the colorant in the syrup at the outlet higher than its specification. The finding of this paper would allow engineers to optimize an industrial-scale pulsed bed adsorber more efficiently.

1. Introduction

Pulsed-bed adsorbers are adsorbers that the the liquor is fed at the bottom of the bed of adsorbent similarly to the fixed-bed but the spent adsorbent is periodically removed from the bottom while the new or on-site regenerated adsorbent is added to the top at the specified time interval so that the height of bed remains unchanged. This type of operation is referred as the pulse operation and the period of the adsorbent removal and the replacement of new adsorbent is called the cycle time [1-3]. This pulse operation simulated the countercurrent movement of spent adsorbent and liquor; hence, the efficiency of mass transfer is greater when comparing with the fixed-bed adsorbers. Usually, the pulsed-bed adsorbers are preferred when the usage rate of adsorbent and the liquor throughput are relatively high. In order to make the pulsed bed adsorbers cost competitive, the spent adsorbent is usually regenerated on-site with a limited feedback control, if any, on the effectiveness of regeneration [2, 4]. Presently, pulsed-bed adsorbers have been used in wastewater treatment and recently been applied in sugar refining where the pulsed bed adsorbers of granulated activated carbon (GAC) are applied to remove the colorants in sugar syrup. Among several operating parameters, cycle time of pulse and percentage of the bed removed is frequently used as the operating parameters for controlling pulsed-bed adsorbers. Generally, 5-15% of adsorbent in the bed is removed and replaced with new adsorbent in

Content from this work may be used under the terms of the Creative Commons Attribution 3.0 licence. Any further distribution of this work must maintain attribution to the author(s) and the title of the work, journal citation and DOI.
Published under licence by IOP Publishing Ltd

order to minimize its effect on hydrodynamics and dynamics of the adsorbers while the cycle time is set at 12 - 24 hours in order to minimize the abrasion of adsorbent [1]. Although the pulsed bed adsorbers, which are widely used in industrials, the information on the effect of cycle time, liquor flow rate and regenation efficiency of adsorbent on the performance of pulsed bed adsorbers is relatively scarce. Consequently, it is the aim of this article to explore the effects of operating parameter on the operation of industrial-sacle pulsed bed adsorbers by developing a mathematical model of industrial-scaled pulsed bed adsorbers and then applying the model to investigate the effect of operating parameters; namely, cycle time and flow rate on the performance of an industrial scale pulse bed adsorber, numerically. It is worthwhile to note that, according to the author's knowledge, there is no research article published in the literature that focus on the mathematical modeling of pulsed-bed adsorbers.

In this article, the adsorption equilibrium constant or K and the overall mass transfer coefficient according to the linear driving force model or k was first evaluated from the breakthrough curve of a laboratory GAC fixed-bed adsorber published in the literature [5]. Secondly, the tortuosity of GAC (τ) was calculated from the overall mass transfer coefficient measured and the correlation for film mass transfer (k_f) and axial dispersion coefficient (D_L) obtained from the fitting of breakthrough curve. Thirdly, the mathematical model of an industrial-scale pulsed bed adsorber was developed and solved based on the assumption of ideal plug flow of adsorbent during the pulse. Finally, the effect of fraction of adsorbent removed and replace during the pulsed operation and liquor flow rate on the pulsed bed adsorber was investigated.

2. Theory and method.

2.1. Breakthrough of fixed-bed Adsorber

Based on the assumptions of linear adsorption isotherm, Ruthven showed that the breakthrough of a fixed-bed adsorber can be correlated from the following equation [6]

$$\frac{C}{C_0}=\frac{1}{2}\operatorname{erfc}\left(\frac{7}{8}\sqrt{\xi}-\frac{7}{8}\sqrt{\tau}\right) \qquad (1)$$

where

$$\tau=k\left(t-\frac{L}{v}\right) \qquad (2)$$

and

$$\xi=\frac{kKz}{v}\frac{\left(1-\varepsilon\right)}{\varepsilon} \qquad (3)$$

C was the concentration at the outlet, C_0 was the feed concentration, t was the time, L was the height of adsorption bed, v was the liquor interstitial velocity, ε was the bed porosity and τ and ξ are dimensionless variables.

2.2. Modeling of pulsed-bed Adsorbers

By modifying the initial conditions of the mathematical model of a fixed-bed adsorber for plug flow and linear driving force model [6], the mathematical model for an industrial-scale pulsed bed adsorber are:

$$\frac{\partial(C/C_0)}{\partial t}=-\left(\frac{1-\varepsilon}{\varepsilon}\right)\frac{\partial(\bar{q}/C_0)}{\partial t}-v\frac{\partial(C/C_0)}{\partial t} \qquad (4)$$

$$\frac{\partial(\bar{q}/C_0)}{\partial t}=kK\left(C/C_0-C_{eq}/C_0\right) \qquad (5)$$

The initial condition at the first cycle are;

$$C(t=0,z)=0 \qquad (6)$$

$$\bar{q}(t=0,z)=0 \qquad (7)$$

$$C(t,z=0)=C_0 \qquad (8)$$

IOP Conf. Series: Materials Science and Engineering **559** (2019) 012019 doi:10.1088/1757-899X/559/1/012019

For the other cycle, the initial condition at the beginning of each cycles are

$$C(t=Nt_{cycle},z)=\begin{cases} 0 \text{ if } z\geq(1-X_{pulse})L \\ C(t=Nt_{cycle},z-X_{pulse}L) \text{ if } z\leq(1-X_{pulse})L \end{cases} \tag{9}$$

$$\bar{q}(t=Nt_{cycle},z)=\begin{cases} 0 \text{ if } z\geq(1-X_{pulse})L \\ \bar{q}(t=Nt_{cycle},z-X_{pulse}L) \text{ if } z\leq(1-X_{pulse})L \end{cases} \tag{10}$$

where k was the overall mass transfer coefficient, K was the adsorption equilibrium constant, $\bar{q}$ was the average concentration in the adsorbent, C_{eq} was the concentration in liquid at equilibrium with $\bar{q}$, N was the number of cycle, t_{cycle} was the cycle time, z was the distance from the inlet, X_{pulse} was the fraction of bed that was removed and replaced by the new adsorbent in each cycle and L was the height of the pulsed bed.

2.3. Numerical analysis of pulsed-bed Adsorbers

By fitting equation (1) with the breakthrough curve of sugar colorants of a fixed-bed Reactor reported by Cortes [5], the adsorption equilibrium constant (K) and the overall mass transfer coefficients were obtained. The fitting was carried on by Solver in Microsoft Excel.

Once k was obtained, the correlation of the external (film) mass transfer coefficient or k_f [6] and axial dispersion coefficient or D_L [7] were used to estimate the pore diffusion of sugar colorants (D_p) and totoursity (τ_p) from [6]

$$D_p = \frac{R_p^2/15\varepsilon_p}{\left[\frac{1}{kK}-\frac{D_L}{v^2}\left(\frac{1-\varepsilon}{\varepsilon}\right)-\frac{R_p}{3k_f}\right]} \tag{11}$$

$$\tau_p = D_p/D_M \tag{12}$$

where R_p was the radius of particle, ε_p was the particle porosity and D_M was the molecular diffusivity of sugar colorants in water. Note that the molecular diffusivity of pullunan [8] was used to estimate the molecular diffusivity of sugar colorants since both are similar in origin and the particle porosity was estimated from the total porosity of bed reported by Cortes [5] assuming the typical value of bed porosity of 0.38 [9].

2.4. Estimating parameters of fixed-bed Adsorber from breakthrough curve

The typical of pulsed bed adsorber with the diameter of 4 m and the height of 15 m was used in the numerical analysis. The typical flow rate of sugar syrup of 87.90 m^3/hr was assumed. The superficial velocity at the flow rate was 0.001943 m/s which was comparable to the experiment run by Cortes [5] that the experimental results was used to estimate the parameters; hence, the mechanism in both cases should be the same. The color of feed syrup was typically 700 ICUMSA and the color of sugar syrup at the outlet was controlled at 100 ICUMSA. The ratio of the color at the outlet to that at the inlet will be used as the ratio of concentration at the outlet to that at the inlet or C/C_0.

The numerical method of lines was used to solve equation (4) and equation (5) with the initial conditions shown in equation (6) to equation (10). The 5-point biased upwinding method was used to approximate the spatial derivative [10] and the resulted ordinary differential equations were solved by ode solver in Scilab which is the open source alternative to Matlab. The step sizes in spatial direction of 0.008442 m and of 0.25 m were appeared to generate the numerical results with acceptable accuracy.

3. Results and discussions

3.1. Fitting of breakthrough curve and evaluate parameters from breakthrough curve

Based on the breakthrough reported by Cortes [5], Equation (1) was used to estimated k and K for the flow rate from 0.023 ml/min to 0.043 ml/min as illustrated in table 1.

Table 1. Adsorption Equilibrium Constant and Overall Mass Transfer Coefficient from Breakthrough Curve.

Flow rate (ml/min)	k (min)	K
23	6.03×10^{-3}	63.41
33	6.35×10^{-3}	63.41
43	8.27×10^{-3}	63.41

From the values of overall mass transfer coefficient, the correlation of k_f [6] and the correlation of D_L [7], the effective diffusivity in porous activated carbon can be evaluated as shown in table 2.

Table 2. Film Mass Transfer Coefficients and Pore Diffusivity of Colorants in Sugar Syrup.

Flow rate (ml/min)	k_f (min)	D_P (m^2/s)
23	8.06×10^{-2}	1.80×10^{-10}
33	9.07×10^{-2}	1.89×10^{-10}
43	9.90×10^{-2}	2.46×10^{-10}

From the average value of D_p and the value of D_M, the value of totuosity of activated carbon was approximated to be 2.012 which was in the range of 2 – 10 reported for activated carbon [4].

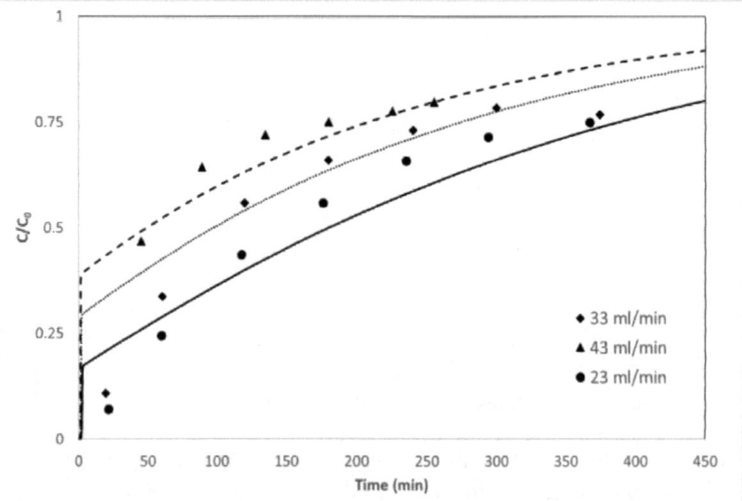

Figure 1. The Experimental and Correlated Breakthrough Curve.

From the values of D_p, D_L and K, the breakthrough curves of sugar colorants from a fixed-bed adsorber at different flow rate was shown in figure 1. From figure 1, it was obvious that the parameters obtained can describe the experimental results adequately especially at 33 and 43 ml/min.

3.2. Numerical analysis of an industrial-scale pulsed-bed Adsorber

From the values of parameters obtained from the experimental results of a fixed-bed adsorber, the breakthrough of a fixed-bed adsorber with the same dimension as a pulsed bed was shown in figure 2 along with the concentration profile in a pulsed bed adsorber in figure 3.

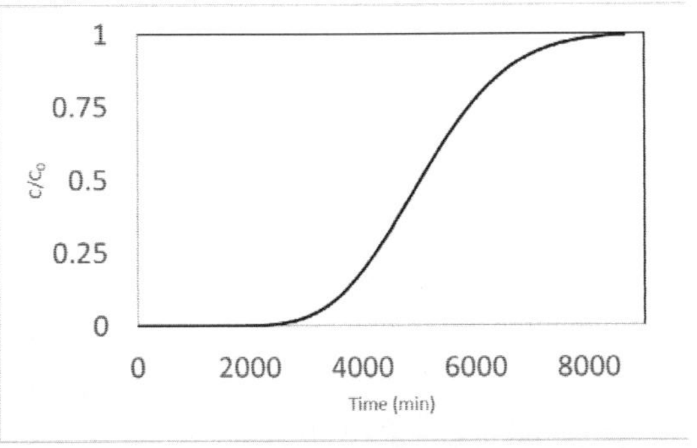

Figure 2. The Predicted Breakthrough Curve from a Fixed-bed Adsorber with the same Size as an Industrial-scale Pulsed Bed Adsorber.

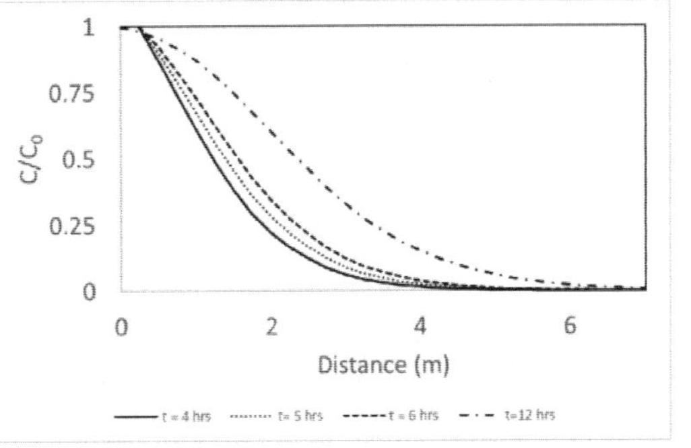

Figure 3. The Concentration Profile in a Fixed-bed Adsorber with the same Size as an Industrial-scale Pulsed Bed Adsorber at different time.

Then, the mathematical model of an industrial-scale pulsed bed adsorber was solved and illustrated as shown in figure 4 to 8. From figure 4 to 8, it can be seen that the value of C/C_0 would reach the pseudo-steady state value after 40 cycles.
Furthermore, if t_{cycle} was less than the optimum cycle time, C/C_0 will be higher than the specification of processed syrup set at 0.1429. The optimum cycle time would depend on X_{pulse}. When X_{pulse} was 0.05, the optimum cycle time was 4 hours and, when X_{pulse} was 0.14, the optimum cycle time was 12 hours. Comparing with figure 3, the utilization of activated carbon will be more efficient if X_{pulse} is less.

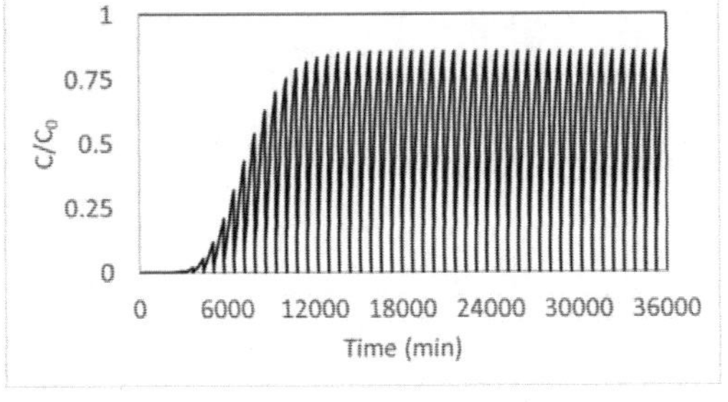

Figure 4. Predicted Performance of an Industrial-scale Pulsed Bed Adsorber for X_{pulse} = 0.05 and t_{cycle} = 12 hours.

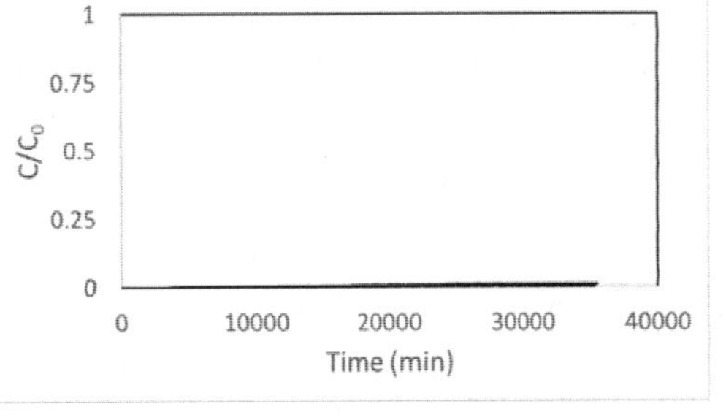

Figure 6. Predicted Performance of an Industrial-scale Pulsed Bed Adsorber for X_{pulse} = 0.05 and t_{cycle} = 4 hours.

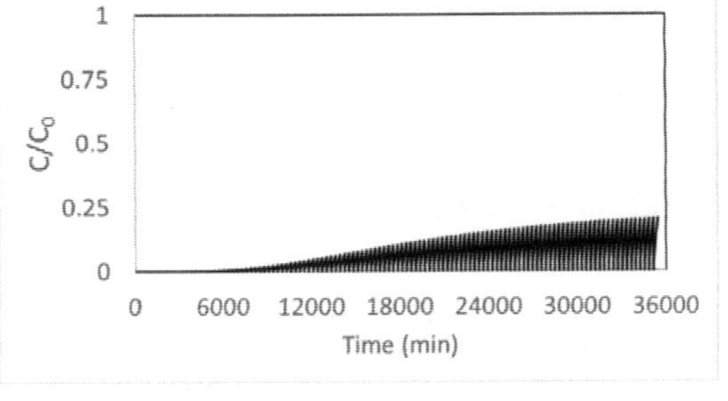

Figure 5. Predicted Performance of an Industrial-scale Pulsed Bed Adsorber for X_{pulse} = 0.05 and t_{cycle} = 6 hours.

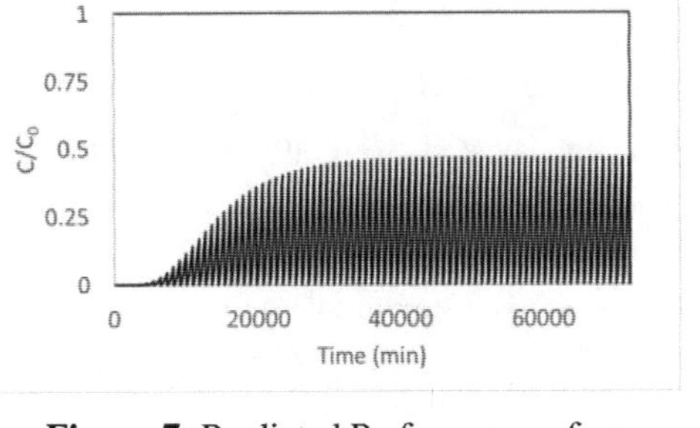

Figure 7. Predicted Performance of an Industrial-scale Pulsed Bed Adsorber for X_{pulse} = 0.14 and t_{cycle} = 15 hours.

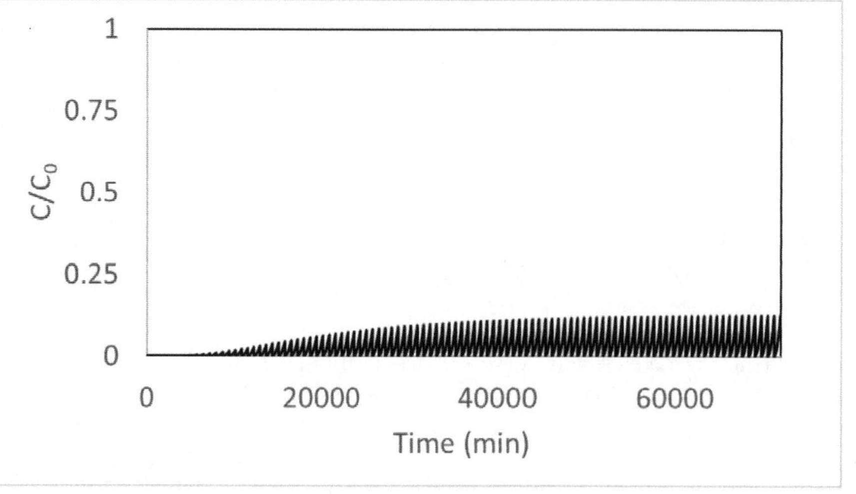

Figure 8. Predicted Performance of an Industrial-scale Pulsed Bed Adsorber for X_{pulse} = 0.14 and t_{cycle} = 12 hours.

Then, the effect of flow rate on the optimum cycle time for this industrial-scale pulsed bed adsorber was investigated. It is assumed that the flow rate varied from 70.32, 79.11, and 87.90 m^3/hr or -20% and -10% from the typical flow rate of 87.90 m^3/hr. The analysis showed that if X_{pulse} was still at 0.05, the optimum cycle time would be changed to 13 hours when the flow rate decreased by 10% and to 15 hours when the flow rate decreased by 20%.

From figure 9 and figure 10, it was obvious that the increase in syrup flow rate would disturb the original pseudosteady state and make the process reach the new pseudosteady state. Consequently, the cycle time set for the industrial-scale pulsed bed adsorber must be adjusted so that it can tolerate the fluctuation of syrup feed flow rate. Alternatively, the feed forward model based predictive control can be implemented so that the utilization of adsorbent is maximized.

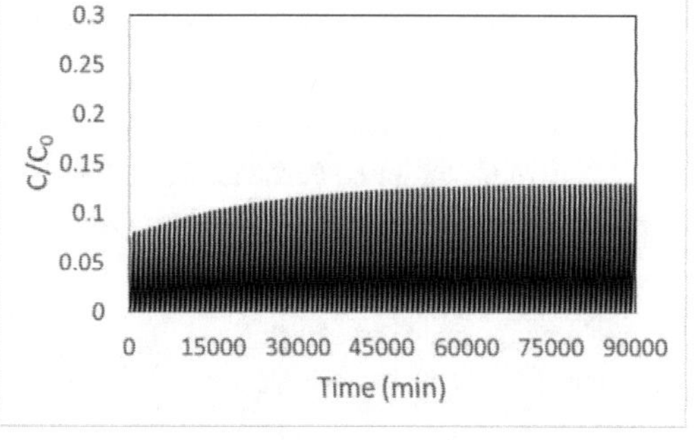

Figure 9. Dynamic Response of an Industrial-scale Pulsed Bed Adsorber for X_{pulse} = 0.14 and t_{cycle} = 15 hours. Originally, the flow rate was set at 70.32 m^3/min; then the flow rate was suddenly increased by 10% for 120 min.

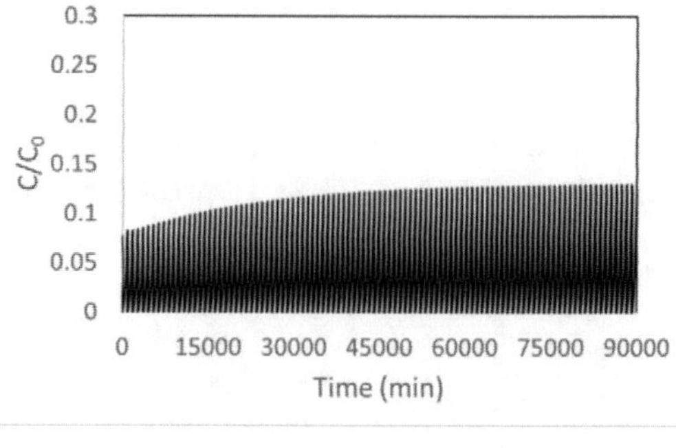

Figure 10. Dynamic Response of an Industrial-scale Pulsed Bed Adsorber for X_{pulse} = 0.14 and t_{cycle} = 15 hours. Originally, the flow rate was set at 70.30 m^3/min; then the flow rate was suddenly increased by 20% for 120 min

4. Conclusions

In conclusion, the mathematical model for an industrial-scale pulsed bed adsorber was developed. By using the data of breakthrough curve of sugar colorant published in the literature, the adsorption and mass transfer parameters of a fixed-bed adsorber were evaluated; the obtained values agreed with the reported ranges in the literature. The parameters were then used in the numerical analysis of an industrial-scale pulsed bed adsorber. In the analysis, it was found that the optimum cycle time depended on the fraction of adsorbent being removed and replaced in each pulse. Generally, the higher the fraction being pulsed, the longer the optimum cycle time. When the fraction being pulsed was kept constant, the decreasing flow rate would yield the longer optimum cycle time. Hence, the feed forward of an industrial-scale pulsed bed on flow rate should maximize the utilization of adsorbent. Further studies on this aspect will be published subsequently.

5. References

[1] C.C. Chou. *Handbook of Sugar Refining A Manual for the Design and Operation of Sugar Refining Facilities*. New York, USA: John Wiley & Sons, 2000.
[2] R.M. Clark and B.W. Jr. Lykins. *Granulated Activated Carbon: Design Operation and Cost*. Michigan, USA: Lewis Publishers, 1989.
[3] J.-H. Liang. Pulsed-bed moving-granular activated carbon system in *Handbook of Sugar Refining A Manual for the Design and Operation of Sugar Refining Facilities* edited by Chou, C.C. New York, USA: John Wiley & Sons, 2000.
[4] M.M. Benjamin and D.F. Lawler. *Water Quality Engineering: Physical / Chemical Treatment Processes*, Wiley, 2013.
[5] R.A. Cortes. *Direct White Sugar Production: Optimization and Chemical Regeneration of Fixed-Bed Activated Carbon Adsorbers*. Master Thesis: Louisiana State University, 2007.
[6] D.M. Ruthven. *Principles of Adsorption and Adsorption Processes*. New York, USA: John Wiley & Sons, 1984.
[7] T.K. Perkins and O.C. Johnston. A Review of Diffusion and Dispersion in Porous Media, *Society of Petroleum Engineers Journal*, pp. 70 – 84, 1973.
[8] S. Matsukawa and T. Brenner. Diffusion measurements of water and polymers in hydrogels by pulsed field gradient NMR in *Nano/Micro Science and Technology in Biorheology*, edited by Kita, R. & Dobashi, T. Tokyo, Japan: Springer, 2015.
[9] Chemviron Carbon. *Canecal Agglomerated Coal Based Granular Activated Carbon*, 2004.
[10] W.E. Schiesser and G.W. Griffiths. *A compendium of Partial Differential Equation Models Method of Lines Analysis with MATLAB*. New York, USA: Cambridge University Press, 2009.

Acknowledgment
The author would like to thank Suranaree University of Technology for financial support.